Active Tectonics of the Devils Mountain Fault and Related Structures, Northern Puget Lowland and Eastern Strait of Juan de Fuca Region, Pacific Northwest

By Samuel Y. Johnson, Shawn V. Dadisman, David C. Mosher, Richard J. Blakely, *and* Jonathan R. Childs

U.S. Geological Survey Professional Paper 1643

Prepared in cooperation with the Geological Survey of Canada

U.S. Department of the Interior
U.S. Geological Survey

U.S. Department of the Interior
Gale A. Norton, Secretary

U.S. Geological Survey
Charles G. Groat, Director

Version 1.0

First printing 2001

For sale by U.S. Geological Survey, Information Services
Box 25286, Federal Center
Denver, CO 80225

Library of Congress Cataloging-in-Publication Data

Active tectonics of the Devils Mountain Fault and related structures, northern Puget Lowland and eastern Strait of Juan de Fuca region, Pacific Northwest / by Samuel Y. Johnson ... [et al.].
p. cm. – (U.S. Geological Survey professional paper; 1643)
Includes bibliographical references.
1. Faults (Geology)—Juan de Fuca, Strait of, Region (B.C. and Wash.)
2. Geology—Juan de Fuca, Strait of, Region (B.C. and Wash.).
I. Johnson, Samuel Y. II. Series.

QE606.5.J83 A37 2002
551.8′09797—dc21 2001033439

ISBN=0-607-97457-5

Contents

Plates

Figures

Tables

Active Tectonics of the Devils Mountain Fault and Related Structures, Northern Puget Lowland and Eastern Strait of Juan de Fuca Region, Pacific Northwest

By **Samuel Y. Johnson, Shawn V. Dadisman, David C. Mosher,[1] Richard J. Blakely, *and* Jonathan R. Childs**

Abstract

Information from marine high-resolution and conventional seismic-reflection surveys, aeromagnetic mapping, coastal exposures of Pleistocene strata, and lithologic logs of water wells is used to assess the active tectonics of the northern Puget Lowland and eastern Strait of Juan de Fuca region of the Pacific Northwest. These data indicate that the Devils Mountain fault and the newly recognized Strawberry Point and Utsalady Point faults are active structures and represent potential earthquake sources.

The north-dipping (45°–75°) Devils Mountain fault extends westward for more than 125 kilometers from the Cascade Range foothills to Vancouver Island. The Devils Mountain fault is bounded by northwest-trending en-echelon folds and faults, a pattern that strongly suggests it is a left-lateral, oblique-slip, transpressional "master fault." Aeromagnetic anomalies coincide with both the trace of the Devils Mountain fault and en-echelon structures. Quaternary strata are deformed on nearly all crossing seismic-reflection profiles, and onshore subsurface data indicate offset of upper Pleistocene strata.

The west-northwest-trending, subvertical Strawberry Point fault cuts across northern Whidbey Island and has a minimum length of about 25 kilometers. On the west coast of Whidbey Island and in the Strait of Juan de Fuca, the fault has south-side-up offset and forms the northern boundary of an uplift of pre-Tertiary basement rock. Exposures and subsurface logs of upper Pleistocene strata from Strawberry Point on the east coast of Whidbey Island indicate that the fault bifurcates into a 2-kilometer-wide zone as it crosses Whidbey Island. Each of the four fault splays within this zone has apparent north-side-up offset, and upper Pleistocene strata between the faults exhibit considerable shortening (dips as steep as 45°). The vertical fault trace, reversal of offset along strike, and evidence for associated contractional deformation suggest that the Strawberry Point fault is an oblique-slip, transpressional fault.

The northwest-trending, subvertical Utsalady Point fault similarly cuts across northern Whidbey Island and has a minimum length of 25 kilometers. It forms the southern margin of the pre-Tertiary basement horst block on the west coast of Whidbey Island, where it has north-side-up offset. Offshore seismic-reflection data from east of Whidbey Island indicate that it also bifurcates eastward into a broad (1.5 kilometer) zone of several splays. Onshore outcrops and subsurface logs from Camano Island indicate a probable reversal of offset (to south side up) along the zone and display both faulting and folding (dips as steep as 24°) in upper Pleistocene strata. As with the Strawberry Point fault, the vertical fault trace(s), reversal of offset, and evidence for associated contractional deformation suggest that the Utsalady Point fault is an oblique-slip, transpressional fault.

Collectively, the Devils Mountain, Strawberry Point, and Utsalady Point faults represent a complex, distributed, transpressional deformation zone. The cumulative slip rate on three main faults of this zone probably exceeds 0.5 millimeter/year and could be much larger. This new information on fault location, length, and slip rate should be incorporated in regional seismic hazard assessments.

Introduction

Seismic hazard assessments depend on understanding the locations, lengths, and slip rates of active faults. Obtaining this information is difficult in the Puget Lowland of Washington State, where most landforms were generated during the last glaciation and are quite young (≈12–20 ka), and vegetation covers much of the landscape. As a result, the earthquake hazard posed by many significant faults in this region is poorly understood. Several potentially significant faults occur in the northern Puget Lowland and eastern Strait of Juan de Fuca region (Gower and others, 1985). These include the southern Whidbey Island fault, the Devils Mountain fault, and associated structures (Gower and others, 1985; Johnson and others, 1996).

Johnson and others (1996) integrated seismic-reflection data, well logs, and information from outcrops to document Quaternary deformation along the southern Whidbey Island fault. Our purpose in this report is to present comparable new geologic and geophysical information and synthesis that demonstrate Quaternary activity on the Devils Mountain fault and two newly defined structures, the Strawberry Point and Utsalady Point faults. These faults are the major components of an oblique-slip transpressional deformation zone that extends westward for more than 125 kilometers from the Cascade Range foothills across the eastern Strait of Juan de Fuca.

Acknowledgments

For the U.S. Geological Survey seismic-reflection surveys, we thank Guy Cochrane, Kevin O'Toole, Larry Kooker, Fred Payne, Curtis Lind, and the captain and crew of the *Robert Gray* for help in data collection, and John Miller and William Stephenson for help in data processing. For the Geological Survey of Canada seismic-reflection surveys, we thank the captain,

[1]Geological Survey of Canada, P.O. Box 1006, Cartmouth, NS CANADA B2Y 4A2.

crew, and scientific staff of the *CCGS John P. Tully* on cruises PGC96006 and PGC97007. The U.S. Geological Survey research cruises were supported by the Coastal and Marine Geology Program and the Earthquake Hazards Reduction Program (EHRP). Geological Survey of Canada surveys were funded in part by a U.S. Geological Survey EHRP grant to David C. Mosher. We thank Michael Fisher for organizing the Seismic Hazard Investigations in Puget Sound (SHIPS) marine seismic-reflection survey and providing access to the SHIPS data base. We are grateful to Mobil and ARCO for providing industry seismic-reflection data. Douglas Kelly and Myrtle Jones provided important water-well data and insights, and Jon Cox drafted early versions of several stratigraphic correlation diagrams. David Dethier, Donald Easterbrook, James Yount, Joseph Dragovich, and David Norman provided important insights on the glacial geology of northern Whidbey Island. Constructive reviews were provided by Thomas Brocher, David Dethier, Donald Easterbrook, Joseph Dragovich, Derek Booth, Kathy Troost, John Oldow, and David Harding. We benefited from stimulating discussions with these reviewers and with Robert Bucknam, James Evans, Carol Finn, Art Frankel, Ralph Haugerud, Robert Karlin, Harvey Kelsey, Daniel Muhs, Alan Nelson, Thomas Pratt, Rowland Tabor, Timothy Walsh, Craig Weaver, Ray Wells, and John Whetten.

Geologic Setting

Western Washington and southwestern British Columbia lie within the structurally complex continental margin of the Pacific Northwest (fig. 1). Oblique convergence of the Juan de Fuca plate below North America (Engebretson and others, 1985), along with larger-scale shearing of the Pacific Plate against North America, provides the driving force for crustal faulting and deformation. (See, for example, Wang, 1996; Wells and others, 1998; Stanley and others, 1999.) The Washington Coast Range and Olympic Mountains represent part of a forearc sliver that is moving northward relative to the Cascade Range and is buttressed to the north by pre-Tertiary basement in the San Juan Islands and on Vancouver Island. The boundary between these two terranes is apparently a distributed deformation zone that extends through Washington's Puget Lowland (Johnson and others, 1996). The active Seattle fault (Bucknam and others, 1992; Johnson and others, 1994, 1999; Nelson and others, 1999) and southern Whidbey Island fault (Johnson and others, 1996) occur within this distributed zone. Both the Seattle fault and the southern Whidbey Island fault are associated with significant aeromagnetic and gravity anomalies (for example, Finn and others, 1991; Blakely, Parsons, and others, 1999; Blakely, Wells, and Weaver, 1999), marking the southern boundary of the Seattle basin and the southwest boundary of the Everett basin (fig. 1), respectively.

Johnson and others (1996) have described the southern Whidbey Island fault (fig. 1) as a broad (6–11 km) transpressional zone comprising three main splays, within which the local late Quaternary uplift rate is at least 0.6 mm/yr. This zone represents part of the boundary of the forearc sliver, juxtaposing Eocene marine basaltic basement on the south and southwest with heterogeneous pre-Tertiary basement on the northeast. This crustal boundary is buried in the Puget Lowland but is exposed on southern Vancouver Island as the Leech River fault (fig. 1; Muller, 1983; Clowes and others, 1987).

The Devils Mountain fault (Hobbs and Pecora, 1941; Loveseth, 1975; Tabor, 1994) forms the northern boundary of the Tertiary to Quaternary Everett basin (Johnson and others, 1996) and is associated with an alignment of aeromagnetic anomalies that extend more than 100 km from the Cascade Range foothills to Vancouver Island (figs. 2, 3). At its east end, the Devils Mountain fault merges with the north-trending Darrington fault zone (Tabor, 1994). In the Cascade Range foothills, the fault forms a prominent $\approx$30-km-long topographic lineament aligned with Lake Cavanaugh (fig. 4). Farther west, the Devils Mountain fault trace extends through Quaternary deposits of the Skagit River delta and Whidbey Island and into the eastern Strait of Juan de Fuca. Based on interpretation of petroleum-industry seismic-reflection data, Johnson and others (1996) suggested that the fault continues its westerly trend across the strait (fig. 1). Previously, Whetten and others (1988) and Tabor (1994) suggested that to the west of Whidbey Island, the trace of the Devils Mountain fault extends northwest into an exposed fault zone that juxtaposes disparate pre-Tertiary rocks in the southern San Juan Islands.

Loveseth (1975) suggested left-lateral slip on the Devils Mountain fault based on its onland linearity, juxtaposition of upper Eocene to lower Oligocene continental and marine strata, and horizontal slickensides (Hobbs and Pecora, 1941) in the fault zone. Whetten (1978) suggested as much as 60 km of left-lateral slip. Tabor (1994) inferred that the Devils Mountain fault was continuous with the north-trending Darrington fault zone to the east (fig. 1) and suggested that pre-34 Ma right-lateral slip was probable, based on fault fabric and offset rock units. Evans and Ristow (1994), who also regarded the Devils Mountain and Darrington faults as one continuous structure, argued for left-lateral slip on the basis of orientations of folds in rocks straddling the fault. Johnson and others (1996) noted that on seismic-reflection data from the eastern Strait of Juan de Fuca, the Devils Mountain fault is locally imaged as a north-dipping reverse fault.

Oligocene rocks are clearly deformed along the Devils Mountain fault zone, but an upper age limit for faulting has not yet been determined. Loveseth (1975) and Adair and others (1989) did not report deformation in upper Pleistocene strata along the fault in the Cascade Range foothills, but both noted the youthful appearance of the topographic lineament coinciding with the structure (fig. 4). Naugler and others (1996) first reported Quaternary displacement on the Devils Mountain fault on the basis of high-resolution seismic-reflection data from Lake Cavanaugh (figs. 1, 4) that reveal both faulting and folding of inferred upper Pleistocene to Holocene lacustrine sediments. In a regional study of latest Pleistocene glaciomarine deposition, Dethier and others (1995, their fig. 7) noted that the Devils Mountain fault coincides with a significant ($\approx$10 m) inflection point on a graph of the maximum altitude of glaciomarine drift.

South of the Devils Mountain fault, Gower and others (1985) proposed the presence of another potentially active structure, the northern Whidbey Island fault, based on interpretation

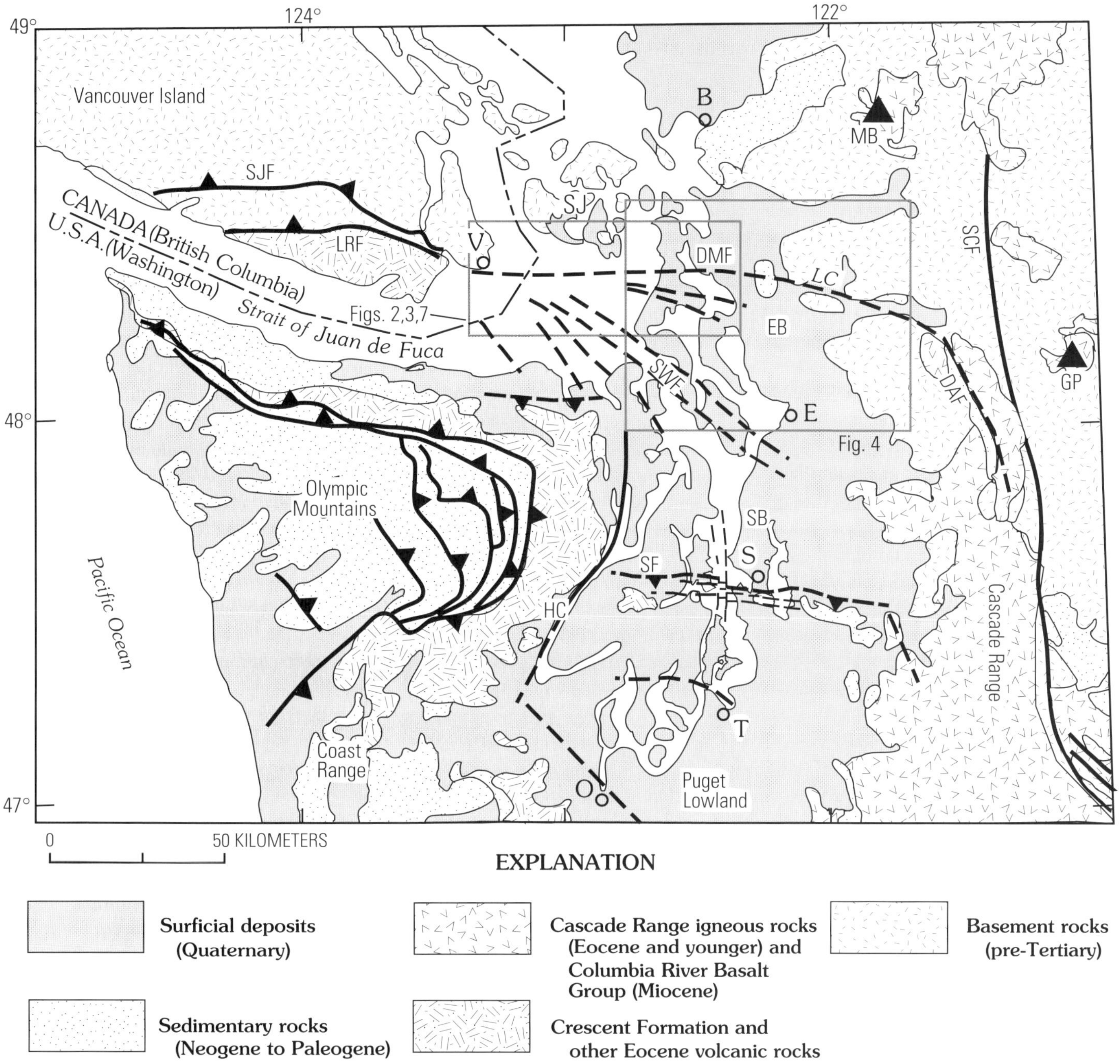

Figure 1. Schematic geology of northwestern Washington and southwestern British Columbia showing Puget Lowland and flanking Cascade Range and Olympic Mountains, and southern Vancouver Island. B, Bellingham; E, Everett; O, Olympia; S, Seattle; T, Tacoma; V, Victoria. Heavy line, fault; dashed where inferred, sawteeth on upper plate of thrust fault. Triangle, modern Cascade volcano. DAF, Darrington fault; DMF, Devils Mountain fault; EB, Everett basin; GP, Glacier Peak; HC, Hood Canal; *LC*, Lake Cavanaugh; LRF, Leech River fault; MB, Mount Baker; SB, Seattle basin; SCF, Straight Creek fault; SF, Seattle fault; SJ, San Juan Islands; SJF, San Juan fault; SWF, southern Whidbey Island fault. Geology from maps and compilations of Tabor and Cady (1978), Washington Public Power Supply System (1981), Gower and others (1985), Walsh and others (1987), Tabor and others (1988, 1993), Whetten and others (1988), Yount and Gower (1991), Tabor (1994), Johnson and others (1996, 1999, 2001).

of gravity and magnetic anomalies (MacLeod and others, 1977), and on faulting of Quaternary sediments exposed at Strawberry Point on northeastern Whidbey Island (fig. 2). Our offshore and onland investigations, including recent high-resolution aeromagnetic surveys (Blakely, Wells, and Weaver, 1999; Blakely and Lowe, 2001), reveal two faults in the vicinity of the inferred northern Whidbey Island fault. Neither of these faults coincides with the trace of the northern Whidbey Island fault as inferred by Gower and others (1985), so we do not use that name in this report and recommend that it be abandoned. Instead, we refer to the two structures as the Strawberry Point fault and the Utsalady Point fault, named for sites (fig. 2) where the faults have deformed upper Pleistocene deposits exposed in coastal bluffs.

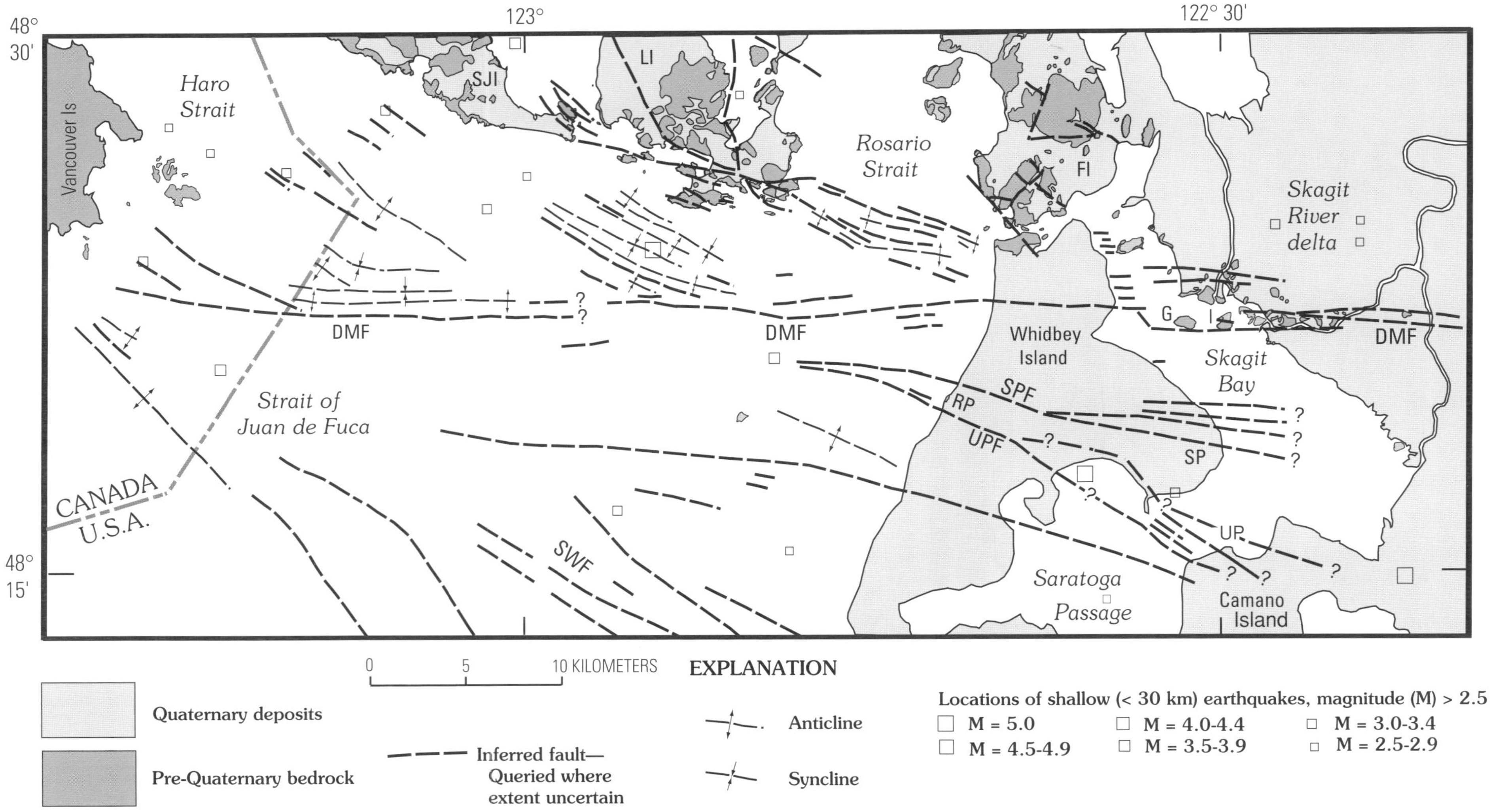

Figure 2. Structural geology of northeastern Strait of Juan de Fuca area (fig. 1), based on interpretation of seismic-reflection (figs. 8-26), aeromagnetic (fig. 3), outcrop (fig. 27), and subsurface data (fig. 28). Onshore faults (except on Whidbey and Camano Islands) are based on mapping of Whetten and others (1988). Not all faults and folds deform Quaternary deposits. Earthquake locations and magnitudes are from the Pacific Northwest Seismograph Network. DMF, Devils Mountain fault; FI, Fidalgo Island; G, Goat Island; I, Ika Island; LI, Lopez Island; RP, Rocky Point; SJI, San Juan Island; SP, Strawberry Point; SPF, Strawberry Point fault; SWF, southern Whidbey Island fault; UP, Utsalady Point; UPF, Utsalady Point fault.

Figure 3. Shaded relief map of the aeromagnetic anomaly map, eastern Strait of Juan de Fuca–northern Puget Lowland region (fig. 1). Illumination is from the northeast. Heavy white lines, coastlines. Light white lines, faults and folds of figure 2. Vertical white line on west part of diagram, boundary between two different magnetic surveys. CI, Camano Island; FI, Fidalgo Island; LI, Lopez Island; RP, Rocky Point; SJF, Strait of Juan de Fuca; SD, Skagit River delta; SJI, San Juan Island; VI, Vancouver Island; WI, Whidbey Island.

Figure 4. SLAR (side-looking airborne radar) image of portion of northwestern Washington (fig. 1). White arrows show topographic lineament associated with Devils Mountain fault. Middle arrow also shows location of elongate Lake Cavanaugh (LC), which occurs along the lineament. CI, Camano Island; DMF, Devils Mountain fault; SD, Skagit River delta; WI, Whidbey Island. Area of image shown in figure 1.

Seismic-Reflection Surveys—Data Acquisition and Processing

High-resolution seismic-reflection surveys (Mosher and Simpkin, 1999) have been shown to be important tools in interpretation of active faults (for example, Barnes and Audru, 1999; Johnson and others, 1999). Our investigation of active structures in the northern Puget Lowland and eastern Strait of Juan de Fuca relies heavily on a network of multichannel seismic-reflection data collected in 1995 by the U.S. Geological Survey (USGS) and in 1996 by the Geological Survey of Canada (G.S.C.). Collectively, we acquired several hundred kilometers of data in the eastern Strait of Juan de Fuca and Skagit Bay (fig. 5, on pl. 1), crossing the Devils Mountain fault 15 times and the Strawberry Point and Utsalady Point faults 6 times.

For the 1995 USGS survey (fig. 5*A*), the seismic source consisted of two 0.655-L (liter) (40 in.^{3}) airguns fired at 12.5-m intervals. Data were digitally recorded for 2 s using a 24-channel (6.25-m group interval; 150-m active length) streamer. Resulting common mid-point (CMP) stacked data are six-fold and have a 3.125-m CMP spacing. Profiles were located using GPS satellite navigation with an accuracy of ±100 m. Relative locations from fix to fix are accurate to within a few meters. These data were deconvolved and filtered before and after stack, then time-migrated using a smoothed velocity function. Deconvolution was only partly effective in suppressing water-bottom multiples.

For the 1996 G.S.C. survey (fig. 5*C*), the seismic source consisted of a 0.655-L (40 in.^{3}) sleeve airgun fired at 16-m intervals, and data were digitally recorded for 1 s using a 24-channel (8-m group interval, 192-m active length) streamer. GPS satellite navigation with an accuracy of ±20 m was used to locate profiles. CMP stacked data are four-fold and have a 4-m CMP spacing.

Both USGS and G.S.C. data are typically of high quality in the upper approximately 0.5–0.7 s and degrade significantly with greater depth. On both the USGS and G.S.C. cruises (fig. 5*A*, *C*), higher frequency seismic-reflection data were collected concurrently with the airgun data. The USGS used a geopulse boomer source and recorded data for 500 ms on a single-channel streamer. The G.S.C. used a Huntec hydrosonde deep tow boomer system and recorded data for 250 ms on a single "internal" hydrophone and a short external streamer trailing behind the towed source. The quality of these higher frequency data is highly variable due to local nonreflective and (or) gassy sub-bottom sediments. Maximum penetration is about 150–200 ms (120–160 m).

The U.S. Geological Survey, the Geological Survey of Canada, and several academic institutions collaborated in 1998 on a deeper multichannel seismic-reflection survey, "Seismic Hazard Investigation of Puget Sound" (SHIPS; Fisher and others, 1999), that traversed the eastern Strait of Juan de Fuca (fig. 5*B*). These data were collected using a 79.3-L (4,838 in.^{3}) airgun array and recorded on a 2.4-km-long, 96-channel streamer. To enhance the shallow (<10 km) geologic section for the purposes of this report, the near-offset 48 channels of these data were stacked (24-fold, 12-m CMP) and velocity migrated. We also examined several proprietary industry seismic-reflection data (3 to 5 s records) profiles from the eastern Strait of Juan de Fuca (fig. 5*B*), most of which were collected in the late 1960's or early 1970's with airgun sources of various sizes, when the area was considered a frontier for petroleum exploration. Examples of these industry data included in this report have been stacked but not migrated.

Stratigraphy and Seismic Stratigraphy

Several distinct stratigraphic units occur onshore in this portion of the northern Puget Lowland (fig. 6). Four different units, distinguished on the basis of stratigraphic position and seismic stratigraphic facies (for example, Sangree and Widmier, 1977; Stoker and others, 1997), are imaged on offshore seismic-reflection data. These include pre-Tertiary basement rocks, Tertiary sedimentary rocks, uppermost Pliocene(?) to Pleistocene strata, and uppermost Pleistocene to Holocene strata.

Pre-Tertiary Basement Rocks

Pre-Tertiary bedrock consisting of the Jurassic Fidalgo Complex, which includes ophiolite and tectonically mixed, variably metamorphosed, sedimentary, volcanic, and plutonic rock, is exposed north of the Devils Mountain fault in the Skagit River delta, on northern Whidbey Island and Fidalgo Island, and in the San Juan Islands (Whetten and others, 1988; Tabor, 1994; fig. 2). Ultramafic and mafic rocks of the Fidalgo ophiolite form notable magnetic highs beneath the Skagit River delta, northernmost Whidbey Island, Fidalgo Island, and the eastern San Juan Islands (figs. 2, 3; Whetten and others, 1980). In contrast, Mesozoic sedimentary rocks commonly form subdued aeromagnetic anomalies, such as on southern San Juan Island and southwestern Lopez Island (fig. 3). One small outcrop of pre-Tertiary rocks is exposed south of the Devils Mountain fault on Whidbey Island at Rocky Point (fig. 2). Pre-Tertiary bedrock, consisting of the Mesozoic Leech River complex and Paleozoic to Mesozoic volcanic, plutonic, and metamorphic rocks, also underlies southern Vancouver Island north of the Leech River fault (fig. 1; Roddick and others, 1979). Pre-Tertiary rocks are generally nonreflective and represent acoustic basement on seismic-reflection profiles.

Tertiary Sedimentary Rocks

Tertiary sedimentary rocks crop out adjacent to the Devils Mountain fault east of northern Whidbey Island within the Skagit River delta and in the Cascade Range foothills (Whetten and others, 1988; fig. 1). Units include the Eocene, nonmarine Chuckanut Formation (Johnson, 1984a; Evans and Ristow, 1994) and the upper Eocene to lower Oligocene marine to marginal marine rocks of Bulson Creek (Marcus, 1980) (fig. 6). These rocks are cut by numerous west- and northwest-trending faults and are gently to steeply folded and locally overturned (Whetten and others, 1988; Dragovich and others, 2000). On industry seismic-reflection profiles, Tertiary strata are characterized by relatively continuous, high-amplitude, parallel to subparallel, moderate frequency reflections and can thereby generally be distinguished from nonreflective pre-Tertiary basement.

Uppermost Pliocene(?) to Pleistocene Strata

Onshore Stratigraphy

Uppermost Pliocene(?) to Pleistocene deposits of the Puget Lowland comprise a stratigraphically complex basin fill of glacial and interglacial deposits that are locally as thick as 1,100 m (Yount and others, 1985; Jones, 1996). Easterbrook (1994a, b) described six glacial drift units, three of which are exposed on Whidbey Island (fig. 6). In the northern Puget Lowland, glacial drift typically consists of till, outwash, and glaciomarine deposits. Interglacial deposits are typically fluvial and deltaic, including peat.

The Double Bluff Drift is the oldest Pleistocene unit recognized on Whidbey Island (Easterbrook, 1968, 1994b). It does not crop out on the northern Whidbey Island area of this investigation, where it is inferred to occur below sea level on the basis of stratigraphic correlation diagrams derived from water-well logs. Deposition of the Double Bluff Drift is thought to have occurred between about 250 and 130 ka (Blunt and others, 1987; Easterbrook, 1994b) and probably coincides approximately with marine isotope stage 6 (for example, Shackleton and others, 1983; McManus and others, 1994; Kukla and others, 1997).

The interglacial alluvial- and delta-plain deposits of the Whidbey Formation (Easterbrook and others, 1967; Easterbrook, 1968, 1969; Heusser and Heusser, 1981) overlie the Double Bluff Drift on Whidbey Island. The Whidbey Formation is locally more than 60 m thick and is widespread in the

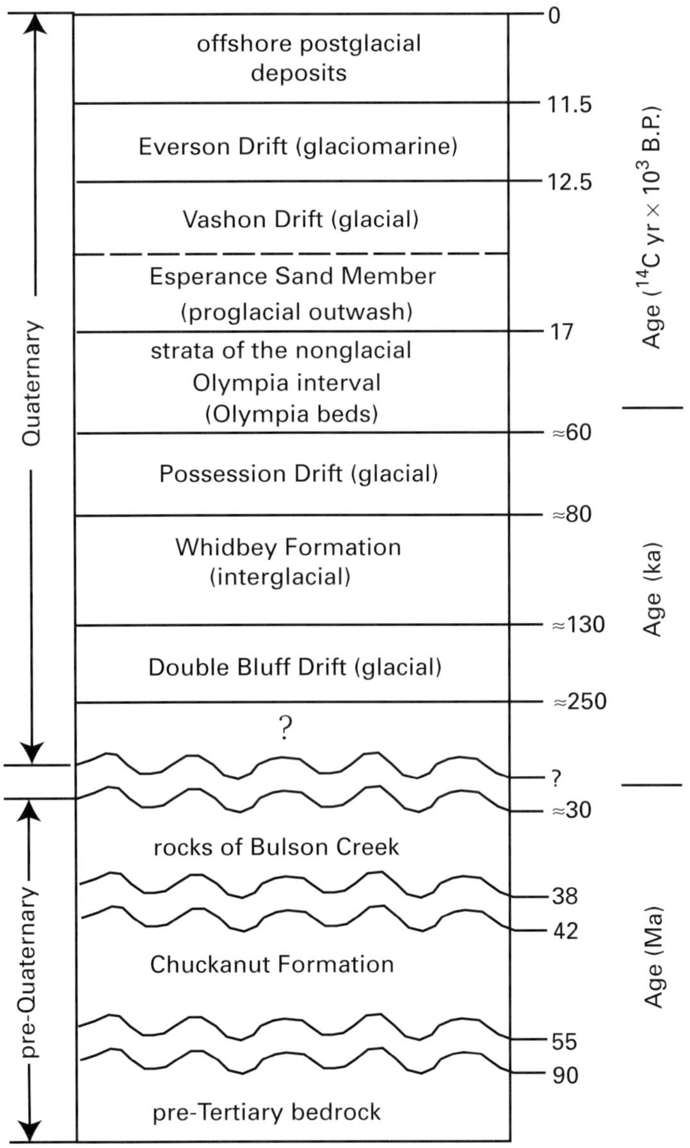

Figure 6. Stratigraphy and age of geologic units occurring in northern Whidbey Island area (figs. 1, 3). Ages for bedrock units based on Johnson (1984a), Whetten and others (1988), and Rau and Johnson (1999). Ages for Quaternary units based mainly on Blunt and others (1987), Berger and Easterbrook (1993), Easterbrook (1994a, b), Porter and Swanson (1998), and Troost (1999).

central and northern Puget Lowland (Easterbrook, 1968, 1994b). Amino-acid analyses of shells and wood at 13 localities suggest an age of 100±20 ka for the Whidbey Formation; four thermoluminescence ages range from 102±38 to 151±43 ka (Blunt and others, 1987; Berger and Easterbrook, 1993). These dates indicate a correlation with interglacial marine isotope stage 5 (for example, Shackleton and others, 1983; McManus and others, 1994; Kukla and others, 1997), and pollen data indicate internal climatic fluctuations occurred during its deposition (Heusser and Heusser, 1981; T.A. Ager, written commun., 2000). We therefore infer that the top of the Whidbey Formation (where it is not significantly eroded) has an age corresponding to the end of this stage (and not the age of a stage 5 substage), about 80 ka. Our investigations suggest that the top of the Whidbey Formation can be recognized and traced in the northern Whidbey Island area using water-well logs (see "Evidence for Faulting Onshore, Whidbey and Camano Islands"), and therefore provides a potential marker for estimating both amounts and rates of recent deformation.

The Possession Drift discontinuously overlies the Whidbey Formation (fig. 6). At its type locality on southern Whidbey Island, it thins laterally through a distance of about 400 m from a thickness of 25 m to an unconformity between the Whidbey Formation and younger sediments (Easterbrook, 1968). Stratigraphic correlation diagrams from northern Whidbey Island based on water-well logs (see "Evidence for Faulting Onshore, Whidbey and Camano Islands") suggest a similarly variable occurrence and thickness for this unit. Radiocarbon dates from the Possession Drift yield infinite ages (>≈45,000 ^{14}C yr B.P.); amino acid analyses of marine shells in Possession Drift at several localities suggest an age of about 90–75 ka (Blunt and others, 1987). Based on these ages and those from underlying and overlying strata, the Possession Drift appears correlative with marine isotope stage 4 (for example, Shackleton and others, 1983; McManus and others, 1994) and was probably deposited between about 80 and 60 ka.

The informally named nonglacial alluvial and deltaic Olympia beds locally overlie Possession Drift or older deposits in the central and northern Puget Lowland (fig. 6). Outcrop and water-well data (see "Evidence for Faulting Onshore, Whidbey and Camano Islands") indicate that these strata have variable thickness (as much as ≈25 m) and are locally absent on northern Whidbey Island. To the northeast in the Swinomish Island–Skagit River delta area, Dragovich and Grisamer (1998) and Dragovich and others (1998) suggested that Olympia beds are as thick as 50 m based on interpretations of subsurface data. Olympia beds were probably deposited between about 60 and 17 ka (Fulton and others, 1976; Blunt and others, 1987; Clague, 1994; Easterbrook, 1994b; Porter and Swanson, 1998; Troost, 1999). Hansen and Easterbrook (1974) and Easterbrook (1976) described a thin sequence of inferred glacial deposits within the Olympia beds at Strawberry Point (fig. 2), which they considered evidence for the ≈34,000–40,000 ^{14}C yr B.P. "Oak Harbor stade of the Possession glaciation." Fulton and others (1976) and Clague (1978) argued, however, that an ice advance at this time was unlikely because of a lack of evidence to the north in southern British Columbia for correlative ice occupation.

Strata associated with the Fraser glaciation on northern Whidbey Island include advance outwash of the Esperance Sand Member of the Vashon Drift, the main body of the Vashon Drift, and recessional deposits of the Everson Drift (Easterbrook, 1968). These units have variable thickness but collectively are commonly 50–100 m thick or more. Radiocarbon dating suggests that this stratigraphic "package" in the northern Whidbey Island area was deposited between about 17,000 and 12,000 ^{14}C yr B.P. (Easterbrook, 1968, 1969; Blunt and others, 1987; Dethier and others, 1995; Porter and Swanson, 1998). Hicock and Armstrong (1981) presented evidence for a ≈21,000 ^{14}C yr B.P. pre-Vashon, early Fraser till-bearing unit (Coquitlam Drift) to the north in southern British Columbia, but ice from this early Fraser stade apparently did not reach the northern Whidbey Island area.

Offshore Stratigraphy

Two distinct seismic units occur above pre-Tertiary or Tertiary "basement" in the eastern Strait of Juan de Fuca. On both industry and higher resolution seismic-reflection data, the lower of these two units consists of seismic facies typical of glacial deposits (Davies and others, 1997). Characteristics include discontinuous, variable-amplitude, parallel, divergent, and hummocky reflections, with common internal truncation, onlap, and offlap of reflections. Based on this seismic facies and on stratigraphic position, this seismic unit is inferred to comprise uppermost Pliocene(?) to Pleistocene deposits (excluding uppermost Pleistocene postglacial deposits). Given the physiography of the eastern Strait of Juan de Fuca, it is likely that much of this unit consists of recessional glaciomarine drift. Present-day seafloor morphology is largely governed by these drift deposits (Hewitt and Mosher, 2001; Mosher and others, 2001). We did not recognize any internal sequences within the inferred uppermost Pliocene(?) to Pleistocene section that could be traced across the region and might correlate with eustatic fluctuations associated with multiple glacial and nonglacial intervals. We attribute this lack of internal stratigraphy to irregular and large-scale glacial erosion and deposition.

Identification and Age of the Base of the Uppermost Pliocene(?) to Pleistocene Section

The base of the inferred uppermost Pliocene(?) to Pleistocene seismic unit is typically most distinct on conventional industry seismic-reflection data (fig. 5*B*), where it is recognized on the basis of contrasts in seismic facies (Johnson and others, 1996, 1999). As described in the preceding paragraph, these mainly Quaternary strata are generally characterized by low- to moderate-amplitude, discontinuous to continuous, irregular hummocky, divergent, and parallel reflections, with common internal truncation, onlap, and offlap. In contrast, pre-Tertiary rocks are nonreflective, and reflections from underlying Tertiary strata have higher amplitude, are more continuous, and are typically parallel to subparallel. Where Tertiary rocks are folded, this contact is generally an angular unconformity that may pass laterally into a disconformity.

On high-resolution seismic-reflection profiles (fig. 5*A*, *C*), this seismic-unit contact and the contrast in seismic facies between older "basement" and mainly Quaternary strata are generally less distinct. For these profiles, the location of the contact is generally based on projection from nearby conventional industry profiles (fig. 5*B*) or on the basis of locally distinct unconformities and onlapping surfaces. Once the contact at the base of the mainly Quaternary section is identified at one or more locations on individual high-resolution profiles, it can generally be traced across the profile based on reflection continuity. Complete regional coverage is accomplished by thus iteratively combining the industry and high-resolution data. Figure 7 is a contour map based on these data that shows the depth to the base of the uppermost Pliocene(?) to Pleistocene section in the northeastern Strait of Juan de Fuca.

Knowing the age of the base of the uppermost Pliocene(?) to Pliocene section in offshore data is important because it provides a potential marker for estimating rates of deformation. Determining this age is, however, problematic. No boreholes have penetrated the submerged section, and multiple pulses of deep subglacial scour and subsequent filling in offshore areas (for example, Booth, 1994) suggest that the age of the surface may vary locally and that correlation with adjacent dated units on land is untenable. For this investigation, we infer that this surface has a maximum age of ≈2 Ma, coinciding with the age of the first glaciation for which there is evidence in the Puget Lowland (Easterbrook, 1994a, b), slightly older than the Pliocene-Pleistocene boundary (Gradstein and Ogg, 1996). However, because of repeated deep glacial erosion, we think it probable that the oldest deposits in this unit imaged on many seismic-reflection profiles are much younger.

Uppermost Pleistocene and Holocene Strata

Variable-amplitude, parallel, and continuous reflections that fill in local basins bounded by Pleistocene bathymetric highs characterize the uppermost seismic unit recognized in the eastern Strait of Juan de Fuca. East of Whidbey Island, these sediments are inferred to be clay and silt derived from the Skagit River (fig. 2). In the eastern Strait of Juan de Fuca where there are no major inputs of postglacial terrestrial sediment, the uppermost Pleistocene to Holocene basin fill probably consists largely of glacial recessional deposits variably reworked by strong tidal currents.

Seismic-Reflection Profiles

A series of seismic-reflection profiles extending from east to west, across the eastern Strait of Juan de Fuca and through Skagit Bay and Saratoga Passage (figs. 2, 5) document the structural style of the Devils Mountain fault and related structures. Selected lines from this data base are shown in figures 8–26 (pls. 1, 2). Velocities used to estimate Quaternary unit thickness and dip, and vertical exaggeration, are based on analysis of well logs and seismic-reflection data from the region (Johnson and others, 1994, 1996; Pratt and others, 1997; Brocher and Ruebel, 1998): uppermost Pleistocene and Holocene strata—1,600 m/s; uppermost Pliocene(?) to Pleistocene strata—1,800 m/s. All high-resolution airgun and industry profiles are plotted at about the same horizontal distance scale and vertical time scale in order to facilitate comparison, and both interpreted and uninterpreted sections are included. To simplify the discussion, the uppermost Pliocene(?) to Pleistocene section and the overlying postglacial uppermost Pleistocene and Holocene section are in some places collectively referred to as "Quaternary."

Faults on seismic-reflection profiles (figs. 8–26) are recognized on the basis of truncated reflections and abrupt changes in reflection dip or seismic facies, characterized by reflection amplitude, frequency, continuity, and geometry (Sangree and Widmier, 1977; Stoker and others, 1997). Fault dip is generally defined in the upper part of a seismic-reflection profile where the record is best, then projected downdip to the base of the profile. Undeformed beds are characterized by flat reflections and no abrupt changes in dip or seismic facies. Faults are

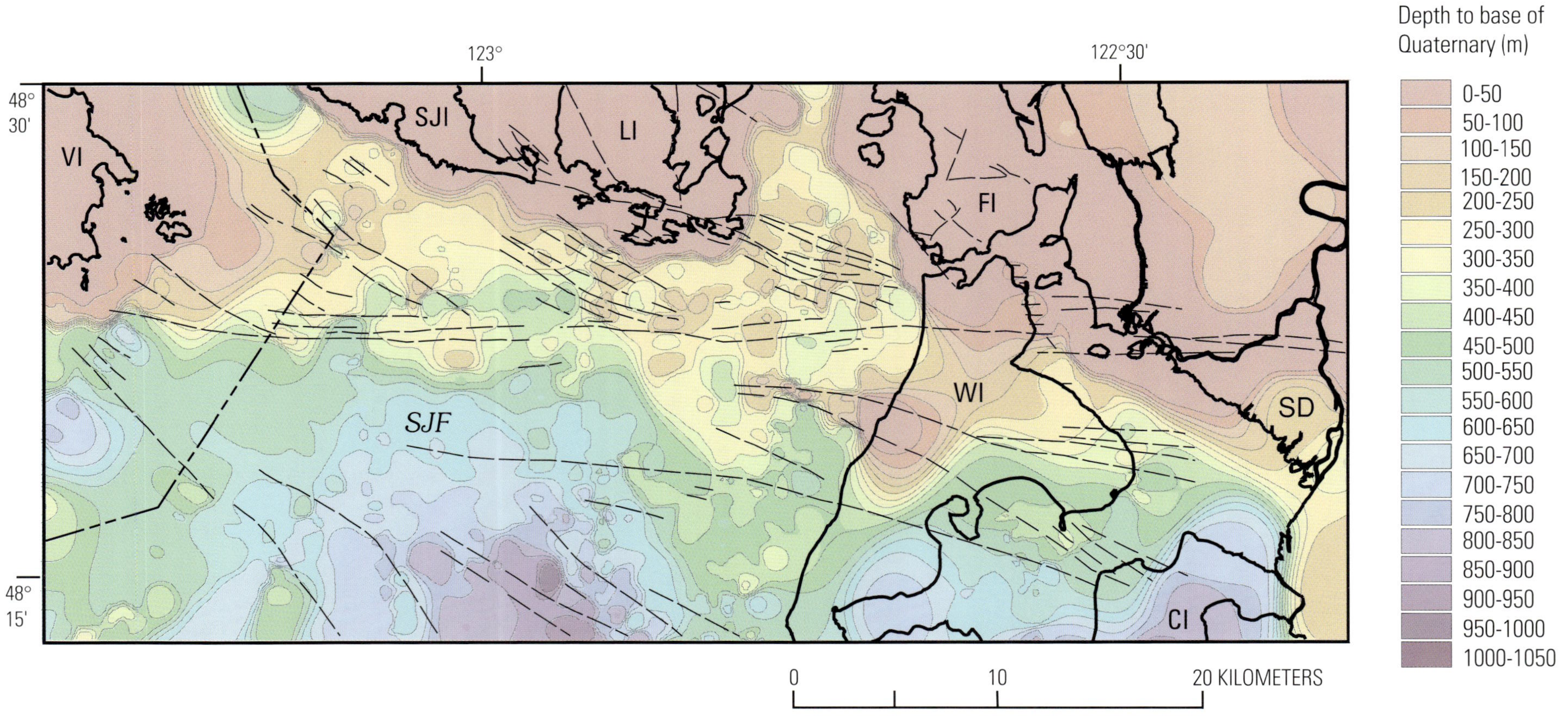

Figure 7. Map showing depth to base of uppermost Pliocene(?) to Pleistocene strata based on outcrops and interpretation of seismic-reflection data, eastern Strait of Juan de Fuca region (see fig. 2). For offshore data points, we converted two-way travel time to depth by assuming velocities of 1,500 m/s for the water column and 1,800 m/s for Quaternary strata. Dark black lines show coastlines. Thin black lines show structures from figure 2. CI, Camano Island; FI, Fidalgo Island; LI, Lopez Island; SJF, Strait of Juan de Fuca; SD, Skagit River delta; SJI, San Juan Island; VI, Vancouver Island; WI, Whidbey Island.

also numbered on the profiles so that text and figures can be easily compared. This numbering scheme is specific to each profile; faults with the same number on different profiles are generally not the same structure.

Aeromagnetic profiles parallel to seismic-reflection tracklines are also shown in figures 8–26. For practical purposes, these profiles have a relatively compressed vertical scale. Thus, they highlight only the largest anomalies; lower amplitude anomalies are more clearly displayed in the figure 3 map.

Seismic-reflection and aeromagnetic data (figs. 3, 5, 8–26) were used together to map the faults and folds across the eastern Strait of Juan de Fuca region shown in figure 2. However, not every structure in figure 2 is shown on every crossing seismic-reflection profile displayed in figures 8–26. All of the structures shown in figure 2 are inferred to cut or warp pre-Tertiary or Tertiary "basement," but many do not obviously deform Quaternary deposits. In some cases, these older structures are deeper than the level imaged by shallower high-resolution seismic-reflection profiles (fig. 5*A*, *C*), and mapping their trends is based entirely on deeper industry data (fig. 5*B*). Conversely, in places the deeper industry data lack the resolution to image structures in the upper 1 km, and mapping of faults and folds is based mostly on the shallower high-resolution profiles.

U.S. Geological Survey Line 176

Line 176 (fig. 8) trends northwest and extends through Skagit Bay on the east side of northern Whidbey Island (fig. 5*A*). On this profile, the combination of shallow water (<50 m), shallow pre-Tertiary basement (inferred from nearby outcrops), and proximity to land has resulted in a seismic record with numerous water-bottom multiples, reverberations, and diffractions. Despite this "noise," numerous faults can be identified on the profile on the basis of truncated and warped reflections, and juxtaposition of panels with disparate seismic facies. Among these faults, the Devils Mountain fault zone is identified based on extension of bedrock faults mapped onland by Whetten and others (1988) 1–2 km east of the profile. The Devils Mountain fault is imaged as a 2-km-wide zone of four subvertical structures. The juxtaposition of disparate seismic facies is largest along the southern fault (6) in the Devils Mountain zone, which correlates with an aeromagnetic low (figs. 3, 8*A*) and is inferred to be the main splay. Both this structure and the fault immediately to the north (7) truncate reflections in the Quaternary section within about 20–30 m (20–30 ms) of the seafloor. At least three other faults are imaged north of the Devils Mountain zone (10, 11, 12), cutting an aeromagnetic high formed by rocks of the Fidalgo ophiolite (Whetten and others, 1988). Of these, only the northernmost fault (12) appears to offset the Quaternary section.

South of the Devils Mountain fault zone, a variably thick (100–250 m; 110–275 ms) uppermost Pliocene(?) to Pleistocene section overlies nonreflective pre-Quaternary strata and is overlain by a relatively thin (≤135 m; 150 ms) uppermost Pleistocene to Holocene (postglacial) section that is largely obscured by water-bottom multiples. At least one fault (5) that extends up into the lower part of the uppermost Pliocene(?) to Pleistocene section occurs in the Everett basin between the Devils Mountain zone and a zone of faults offshore of Strawberry Point.

The structural zone offshore from Strawberry Point consists of at least four subvertical faults (1, 2, 3, 4) that span a distance of 3 km along Line 176, about 2 km perpendicular to the inferred strike of the zone (fig. 2). Although water-bottom multiples are prominent in the upper part of the record in this zone, all four structures appear to propagate upward into the Quaternary section and within about 45–90 m (50–100 ms) of the seafloor. Fault 1 juxtaposes discontinuous, moderate-amplitude, warped reflections (on the southeast) and a largely nonreflective domain, probably cored by uplifted bedrock, to the northeast. Faults 2 and 3 truncate low-amplitude reflections, most obvious between ≈300 and 400 ms. Fault 4 juxtaposes disparate domains of subhorizontal reflections. Overall, uplift of basement rock within and north of the fault zone is suggested by the juxtaposition across fault 1 of relatively nonreflective material to the north and more reflective material to the southeast. No obvious aeromagnetic anomaly is associated with this fault zone along the profile (figs. 3, 8*A*).

U.S. Geological Survey Line 177

Line 177 (fig. 9) trends northeast and extends through Saratoga Passage between eastern Whidbey Island and northern Camano Island (fig. 5*A*). On this line, a 2.3 km-wide fault zone (2 km normal to the inferred fault trend; fig. 2) lies offshore from Utsalady Point. This zone comprises five faults (1 to 5), all of which extend upward into at least the lower part of the Quaternary section. The faults warp, tilt, and truncate reflections, but vertical offset does not appear to be significant across individual faults within the zone or across the fault zone as a whole. Northeast and southwest of this zone, subhorizontal to locally hummocky reflections characterize the Quaternary strata of the Everett basin. Southwest of the zone, significant erosional relief occurs at the base of the inferred postglacial uppermost Pleistocene to Holocene section. The fault zone occurs along the southwest margin of a gentle aeromagnetic high (figs. 3, 9*A*).

U.S. Geological Survey Line 167–168

Line 167-168 (figs. 10, 11, 12) trends north and extends from the eastern Strait of Juan de Fuca into southern Rosario Strait, from about 1 to 5 km west of Whidbey Island (fig. 5*A*). On the airgun profile (fig. 10), the Devils Mountain fault is imaged as a zone of three structures (faults 4, 5, and 6) about 1.5 km wide that correlate with a gentle aeromagnetic high (figs. 3, 10*A*). The northernmost structure (6) is subvertical and is inferred to offset the Tertiary-Quaternary contact about 100–120 m (110–130 ms). This fault juxtaposes a panel of gently south dipping, moderate-amplitude continuous reflections on the north and more discontinuous, variable-amplitude, higher frequency reflections on the south. The fault is overlain by a break in slope in the seafloor, and higher resolution geopulse data (fig. 11) show a contrast in the reflectivity of postglacial (uppermost

Pleistocene to Holocene) sediments coinciding with this break. This contrast indicates juxtaposition of sediments with different physical properties, such as grain size, bedding, and (or) gas-filled porosity (Hovland and Judd, 1988; Fader, 1997). Thus, the fault may project to the surface. The geometry of the break in slope (south side up) is opposite to the inferred sense of vertical slip on the fault at depth, suggesting a possible reversal of slip. Alternatively, uplift could have exposed more erodible sediments on the uplifted northern block, leading to topographic inversion.

The two southern faults in the Devils Mountain fault zone on this profile (4, 5) are subvertical, truncate subhorizontal Quaternary reflections, juxtapose domains of different reflection properties, and propagate upward into at least the lower part of the postglacial uppermost Pleistocene to Holocene section. The amount of vertical separation on the base of the Quaternary across these two structures appears minimal. The juxtaposition of different seismic facies with relatively small vertical offset on both these faults and on fault 4 to the north suggests out-of-plane (strike-slip) movement on these structures.

Two kilometers north of the Devils Mountain fault zone, another steep north-dipping fault (7) truncates reflections on the south-dipping limb of a gentle anticline. The fault appears to project into the lowest part of the uppermost Pliocene(?) to Pleistocene section, and Quaternary strata on the south-dipping limb of the anticline dip south and thin toward the fold axis, indicating deposition during fold growth. The north limb of the fold is also cut by a steep fault (8) that does not appear to extend into the inferred Quaternary section. Northwest-trending structures that occur north of fault 8 in Rosario Strait (fig. 3) mainly affect basement rocks and are not clearly imaged on line 167-168 (fig. 10). The locations of these structures are based on industry data, including profiles collected by Puget Power (1979).

Three faults (1, 2, and 3) occur south of the Devils Mountain fault zone. Based on their westerly traces determined from seismic-reflection data farther west (figs. 10, 12, 13, 14, 15) and on the pattern of aeromagnetic anomalies (fig. 3), these structures are correlated and inferred to be continuous with the faults on the east side of Whidbey Island offshore of Strawberry Point and Utsalady Point (figs. 8, 9), locations for which these faults are here named. The subvertical Strawberry Point fault (3) juxtaposes north-dipping to subhorizontal reflections representing Tertiary and Quaternary strata on the north (largely obscured by multiples) and seismically opaque, uplifted basement rocks and a thin overlying sedimentary section including a small ($\approx$ 850 m wide) postglacial basin on the south. The uppermost Pliocene(?) to Pleistocene section north of and adjacent to the fault dips north about 10°, and vertical separation on the base of this unit across the fault is inferred to be about 80 m ($\approx$90 ms). The companion geopulse profile (fig. 12) shows disrupted reflections in the upper $\approx$10 ms along the fault, but the underlying record is nonreflective.

The subvertical Utsalady Point fault forms the southern margin of the uplift bounded on the north by the Strawberry Point fault and is imaged by at least two splays (1 and 2; fig. 10). The northern splay (within the uplift, fault 2) warps and truncates reflections, juxtaposing reflective and nonreflective zones. The southern splay juxtaposes the uplifted block and subhorizontal reflections in the Everett basin to the south. The faults project upward into at least the lower part of the uppermost Pliocene(?) to Pleistocene section, and vertical offset on the base of this unit across these two splays is about 200–225 m (220–250 ms). The companion geopulse profile (fig. 12) shows these two faults truncating reflections in the inferred uppermost Pleistocene to Holocene section to within 5–10 m of the seafloor. Between the Strawberry Point and Utsalady Point faults, the geopulse profile (fig. 12) also displays another subvertical fault that cuts and warps reflections within both the uppermost Pliocene(?) to Pleistocene and the postglacial uppermost Pleistocene to Holocene sections, to within about 15 m of the seafloor.

The shallow basement between the Strawberry Point and Utsalady Point faults is inferred to consist of pre-Tertiary sedimentary basement rock because of its nonreflective character and because pre-Tertiary sedimentary rock is exposed 1 km to the east at Rocky Point on the west coast of Whidbey Island (fig. 2). The Rocky Point exposures of these basement rocks have surprisingly low bulk magnetic susceptibility (see "Evidence for Onshore Faulting, Whidbey and Camano Islands") and form a prominent aeromagnetic anomaly that extends eastward and provides an important constraint for projection of the Strawberry Point and Utsalady Point faults across Whidbey Island.

U.S. Geological Survey Line 166

Line 166 (fig. 13) trends north and extends from the eastern Strait of Juan de Fuca into southern Rosario Strait, about 6 km west of Whidbey Island (fig. 5*A*). On this profile, the Devils Mountain fault is imaged as two splays (4 and 5) that dip steeply ($\approx$60°–80°) to the north. The southern splay (4) truncates south-dipping reflections within the uppermost Pliocene(?) to Pleistocene section on the south limb of a hanging-wall anticline, as well as nearly horizontal reflections in the Everett basin to the south. The northern splay (5) truncates reflections within the Tertiary section on the south-dipping limb of this fold, but does not obviously break into the Quaternary section. Dips in uppermost Pliocene(?) to Pleistocene beds on the south and north flanks of this fold are about 11° and 7°, respectively. There appears to be about 200–225 m (220–250 ms) of structural relief on the base of the uppermost Pliocene(?) to Pleistocene from the crest of the hanging-wall anticline across the fault zone to the floor of the Everett basin. This inference and estimates of offset on the Devils Mountain fault on this profile are speculative because reflections in the Everett basin are partly obscured by water-bottom multiples. These multiples also partly mask hummocky reflections in the uppermost Pliocene(?) to Pleistocene section south of the Devils Mountain fault, which either could be depositional in origin (such as moraines) or could represent disruption by faulting or folding. A small aeromagnetic anomaly is associated with the Devils Mountain fault on this profile (figs. 3, 13*A*).

Farther south, the Strawberry Point (3) and Utsalady Point (2) faults are imaged as subvertical structures that bound a nonreflective horst block and truncate moderate- to high-amplitude, subparallel, continuous to discontinuous reflections in the

Everett basin. This horst block coincides with an aeromagnetic low (figs. 3, 13*A*) and is inferred to consist of pre-Tertiary sedimentary rock overlain by a thin uppermost Pleistocene to Holocene (postglacial) section. Juxtaposition of this horst block and Everett basin Tertiary and Quaternary strata requires significant uplift. This seismic profile suggests about 150–200 m (165–220 ms) of vertical separation on the base of the uppermost Pliocene(?) to Pleistocene section across the Strawberry Point fault, and about 250–300 m (275–330 ms) across the Utsalady Point fault.

Industry Line 1

Industry Line 1 (fig. 14) is subparallel to U.S. Geological Survey Line 166 (fig. 5*A*, *B*) and provides a slightly deeper view of structure in this area. The Devils Mountain fault is not well imaged, but it appears to juxtapose relatively nonreflective material north of the fault and a section characterized by warped discontinuous reflections to the south (best seen at ≈0.7 s). As on Line 166 (fig. 13), the Strawberry Point and Utsalady Point faults appear as subvertical faults bounding a generally nonreflective uplift that correlates with an aeromagnetic low (fig. 14*A*). The uplift has relief above the seafloor and is onlapped or draped by Quaternary strata, which are warped across the axis of the uplift.

U.S. Geological Survey Line 165

Line 165 (fig. 15) trends west, extending partly across the opening to Rosario Strait, between Lopez and Fidalgo Islands (fig. 5*A*). This line images a gentle asymmetric anticline with a subvertical fault or faults along its axis, coinciding with a bathymetric rise and east-facing scarp on the seafloor. The fault propagates upward into the lower part of the uppermost Pliocene(?) to Pleistocene section, and these strata are involved in the folding. Because water-bottom multiples obscure much of the uppermost part of the record, the amount of deformation in the upper part of the Quaternary section is unclear.

U.S. Geological Survey Line 164

Line 164 (figs. 16, 17) trends north and extends from the eastern Strait of Juan de Fuca into southern Rosario Strait (fig. 5*A*). On the airgun profile (fig. 16), the contact between the inferred Tertiary and uppermost Pliocene(?) to Pleistocene sections is characterized by an upward decrease in frequency, amplitude, and continuity of reflections. This contact is offset about 45–60 m (50–65 ms) across the steeply (≈70°–75°) north-dipping Devils Mountain fault (3), which correlates with a gentle aeromagnetic high (figs. 3, 16*A*). The offset contact is most clear where disparate seismic facies are juxtaposed between about 100 and 300 ms. The companion higher frequency seismic-reflection profile (fig. 17) similarly shows an abrupt vertical termination of reflections at this location, extending up to within 10 m of the surface. This abrupt termination likely indicates a significant contrast in the physical properties of the sediment, such as grain size, stratification, or gas-filled porosity (Hovland and Judd, 1988; Fader, 1997). The steep fault dips and the contrasts in reflection properties across the fault on both images is consistent with out-of-plane (strike-slip) displacements.

Farther south on Line 164 (fig. 16), two faults are mapped as the Strawberry Point (2) and Utsalady Point (1) faults. Both structures are imaged as high-angle faults that truncate subhorizontal reflections, but only fault 1 appears to continue upward into the lower part of the uppermost Pliocene(?) to Pleistocene section. These faults are not recognized on seismic-reflection profiles to the west (for example, figs. 18, 19, pl. 2), and their more subtle seismic and aeromagnetic character (figs. 3, 16*A*) provides an indication that they are dying out (fig. 2). South of these faults, the inferred Tertiary section is warped into a gentle anticline and syncline.

U.S. Geological Survey Line 162

Line 162 (figs. 18, 19) is a north-trending line south of Lopez Island (fig. 5*A*). The airgun line (fig. 18) shows three faults (1, 2, 3) in a 1,200-m-wide zone representing the Devils Mountain fault. The faults dip 60°–80° to the north in the upper ≈300–500 m. On reflections within the Quaternary section across 1, structural relief amounts to about 150–200 m (160–220 ms), and we attribute the bathymetric relief (50–60 m) within the fault zone to Quaternary uplift. All three faults extend upward into the lower part of the inferred uppermost Pliocene(?) to Pleistocene section. The southern strand (1) juxtaposes relatively flat lying beds in the Everett basin and a panel of south-dipping beds in the fault zone. The southerly dips extend upward into the uppermost Pleistocene to Holocene section, indicating that some structural and topographic growth has occurred in the last ≈13,000 years. The middle fault (2) juxtaposes the domain of south-dipping beds to the south with a zone of less reflective north-dipping beds and represents a faulted anticline axis. The northern fault (3) occurs within the north limb of this fold, truncates and warps reflections on its trace, and forms a boundary between the less reflective horizon to the south and a panel of more continuous higher amplitude reflections to the north. Dips north of this fault are about 10°–15° in the uppermost Tertiary section and about 5°–8° in the middle part of the uppermost Pliocene(?) to Pleistocene section. A small positive aeromagnetic anomaly is associated with the fault zone on this profile (figs. 3, 18*A*).

The companion higher frequency profile (fig. 19) provides additional details from the uppermost part of the section. On this profile, the inferred postglacial uppermost Pleistocene to Holocene section is represented by relatively continuous, high-amplitude, high-frequency, parallel reflections; the underlying Pleistocene section is characterized by nonreflective or low-amplitude, discontinuous reflections. The southern strand of the Devils Mountain fault (1 in fig. 18) appears as two splays (1A, 1B). These faults truncate and warp reflections in the upper part of the uppermost Pliocene(?) to Pleistocene section (at ≈170 ms) and warp but do not obviously cut the uppermost Pleistocene to Holocene section. Fault 1B forms the boundary between parallel reflections on the south and a region of warped reflections (dips as much as 8°) and diffractions on the north. The middle fault (2) is not obvious—its location in figure 19 is based on the companion airgun profile (fig. 18). The inferred postglacial uppermost Pleistocene to Holocene section between faults 2 and 3 to the north is very gently warped. Fault 3 truncates and warps the upper part of the uppermost Pliocene(?) to Pleistocene section and cuts the lower and middle parts of the inferred postglacial section to within about 6–7 m of the seafloor (above the stratigraphic level shown in fig. 18).

U.S. Geological Survey Line 158

North-trending Line 158 (fig. 20) is located south of southwestern Lopez Island (fig. 5*A*). The line shows three faults (1, 2, 3) dipping about 45° to 60° to the north (above ≈450 ms). This 1,200-m-wide zone of faults is located about 1,000 m north of the projected westerly trace of the Devils Mountain fault, where no obvious fault is imaged in the upper ≈400 m (≈450 ms). Industry Line 2 (fig. 21*A*, see next section), located 1.5 km to the west of line 158 (fig. 5), reveals "blind" faulting below about 1 km along the projected fault trace, the basis for mapping the fault in this area in figure 2. Faults 1, 2, and 3 all extend upward into the uppermost Pliocene(?) to Pleistocene section, and we infer about 50–100 m (55–110 ms) of structural relief on the base of this unit within the fault zone. Faults 1 and 2 bound a panel of steep north-dipping reflections in basement (inferred Tertiary sedimentary rocks). Fault 3 juxtaposes north- and south-dipping panels and represents a faulted anticline axis. Farther north, the basement and at least the lower part of the uppermost Pliocene(?) to Pleistocene section are warped into a series of gentle anticlines and synclines that have wavelengths of 1–2 km. South of the projection of the Devils Mountain fault, the contact at the base of the uppermost Pliocene(?) to Pleistocene section in the Everett basin is picked mainly on the basis of nearby industry data and is nearly flat. Figure 3 shows that the Devils Mountain fault zone and the fold belt on its north flank correlate with low-amplitude aeromagnetic anomalies.

Industry Line 2

North-trending Industry Line 2 (fig. 21) lies about 1 km west of U.S. Geological Survey Line 158 (fig. 5*A*, B) and provides a deeper image of structure south of Lopez Island. In contrast to the shallower data of figure 20, this profile shows two north-dipping faults (1, 2) along the projected westerly trace of the Devils Mountain fault. Each occurs below about 0.5 s, deeper than the level imaged in figure 20, and neither of these faults obviously disrupts the Quaternary section. Faults 1 and 2 form the structural boundary between subhorizontal beds in the Everett basin to the south and gently folded Tertiary and Quaternary strata in an uplifted block to the north. This broad boundary is overlain by a gentle aeromagnetic high (figs. 3, 21*A*). The folds on the uplifted northern block are cut by at least one north-dipping thrust fault (3) overlain by a hanging-wall anticline. This folded terrane coincides with the southeast portion of a band of aeromagnetic anomalies that extend northwest along the west coast of San Juan Island (fig. 3).

Geological Survey of Canada Line 15

Geological Survey of Canada Line 15 (fig. 22) extends southwest of southern Lopez Island into the eastern Strait of Juan de Fuca (fig. 5*C*). The southernmost fault identified on this profile (1) lies on strike with the westerly projection of the Devils Mountain fault and is inferred to be its main strand. This structure dips about 50°–55° to the north and juxtaposes subhorizontal reflections in Tertiary and uppermost Pliocene(?) to Pleistocene strata (to the south) and a panel of north-dipping reflections in correlative beds to the north. The north-dipping reflections are in turn truncated by another fault (2) about 700 m north of the main strand. Both faults extend upward into the lower part of the uppermost Pliocene(?) to Pleistocene section, and fault 1 appears to offset the base of the postglacial uppermost Pleistocene to Holocene section. South of fault 1, the contact at the base of the uppermost Pliocene(?) to Pleistocene section is a locally angular unconformity.

About 3,600 m north of the inferred main strand of the Devils Mountain fault, another north-dipping fault (3) truncates the south limb of an asymmetric anticline. The fault extends upward into the Quaternary section, and about 100 m (110 ms) of structural relief is present on the inferred base of the uppermost Pliocene(?) to Pleistocene section across the structure. North of this fault, inferred Tertiary and lower Quaternary strata are gently folded and Tertiary strata are cut by at least one fault (4).

Ships Line JDF4

SHIPS Line JDF4 (fig. 23) extends south from offshore southern San Juan Island (fig. 5*B*). This profile images the Devils Mountain fault as a north-dipping (≈45°–55°?) blind structure that truncates and juxtaposes different seismic facies in inferred pre-Quaternary strata. This fault forms the core of an asymmetric anticline that coincides with an aeromagnetic high (figs. 3, 23*A*). The prominent reflection at the base of the uppermost Pliocene(?) to Pleistocene section is folded above the fault, but not ruptured. The dip on the base of this unit on the south limb of the fold is ≈11°, and about 400–450 m (450–500 ms) of inferred structural relief is present on this reflection from the crest of the fold north of the blind fault to the floor of the basin south of the blind fault. Reflections in inferred uppermost Pliocene(?) to Pleistocene strata on the south-dipping limb of the fold dip south and converge toward the anticline axis, suggesting that they were deposited during fold growth. The aeromagnetic high at the south end of the profile probably reflects a change in the lithology of basement rocks, possibly associated with the northwest extent of the southern Whidbey Island fault (Johnson and others, 1996).

Industry Line 3

Industry Line 3 (fig. 24) extends south from Rosario Strait into the eastern Strait of Juan de Fuca (fig. 5*C*). This profile images the Devils Mountain fault as a north-dipping structure (≈60°–70°) that truncates the south limb of an asymmetric anticline. Quaternary beds are gently folded (dips as much as 10°) on the flanks of the anticline but do not appear to be truncated by the fault. The anticline is overlain by an aeromagnetic high (figs. 3, 24*A*) bounded on the south by a steep, west-trending gradient that correlates with the Devils Mountain fault and on the north by a steep, northwest-trending gradient that correlates with a belt of northwest-trending folds.

Geological Survey of Canada Line 37

Geological Survey of Canada Line 37 (fig. 25) extends southwest from offshore of San Juan Island into the eastern Strait of Juan de Fuca (fig. 5*C*). On this profile, the Devils Mountain fault is a north-dipping blind fault (1) overlain by an asymmetric hanging-wall anticline (fault-propagation fold of Suppe and Medwedeff, 1990). Reflections from uppermost Pliocene(?) to Pleistocene beds above the fault on the south-dipping limb of this anticline appear continuous, dip about 7°, and shallow upward. There is approximately 180 m of structural relief on the base of this unit from the crest of this anticline into the flanking footwall basin. Following the methods of Schneider and others (1996, their fig. 10), the horizontal component of fold growth was determined to be about 135 m, and dip on the blind fault is inferred to be about 53°. Lowest uppermost Pliocene(?) to Pleistocene beds also appear to be very gently warped in the syncline and anticline that occur within 2 km north of the Devils Mountain fault, but they flatten upwards. For nearly 6 km north of the Devils Mountain fault, the Quaternary section has a relatively constant thickness of about 200 m (220 ms). Farther north, the basement appears irregular and faulted (faults 2, 3, and others?) and the thickness of the Quaternary section is highly variable.

The uplift on the north side of the Devils Mountain fault forms a prominent aeromagnetic high bounded by west- and northwest-trending gradients on its south and north sides, respectively (figs. 3, 25*A*). The steep west-trending gradient is associated with the Devils Mountain fault. The steep northwest-trending gradient occurs within folded terrane north of the Devils Mountain fault, does not correlate with any feature on the shallow seismic-reflection data, and therefore must have a deeper source. The aeromagnetic high at the south end of the profile similarly has a deeper source and correlates with a significant contrast in basement lithology (Johnson and others, 1996).

Geological Survey of Canada Line 35

Geological Survey of Canada Line 35 (fig. 26) extends southwest from about 4 km east of Vancouver Island into the eastern Strait of Juan de Fuca (fig. 5*C*). On this line, our favored interpretation (Interpretation 1, fig. 26*A*) is that the Devils Mountain fault (1) continues its west trend as a north-dipping (≈45°–55°) structure that offsets the contact between pre-Tertiary rock and uppermost Pliocene(?) to Pleistocene deposits (marked by a set of two high-amplitude reflections) about 250 m (280 ms). Alternatively (Interpretation 2; fig. 26*B*) and less likely, the Devils Mountain fault is a blind structure on this profile (as in the nearest profiles to the east, figs. 23–25), and the north-dipping limb of the fold is poorly imaged on this profile. In this second scenario, Quaternary structural relief is about 500 m (550 ms), Quaternary shortening is about 450–500 m, and fault dip is 45°–50°.

Immediately north of the fault in the hanging wall (Interpretation 1), the pre-Tertiary basement surface dips gently south for about 1,200 m and is draped by uppermost Pleistocene to Holocene (postglacial) beds. Farther north, the pre-Tertiary basement surface is irregular, appears faulted (2 and 3), and is variably overlain by both uppermost Pliocene(?) to Pleistocene and uppermost Pleistocene to Holocene strata. Fault 3 coincides with a steep, continuous, northwest-trending aeromagnetic gradient (figs. 3, 26*A*) and probably represents a significant splay of the Devils Mountain fault.

South of the Devils Mountain fault, uppermost Pliocene(?) to Pleistocene sediment is relatively flat and fills the irregular faulted and folded surface at the top of highly magnetic basement (figs. 3, 26*A*). The prominent anomaly formed by this magnetic basement extends southeast onto the northeastern Quimper Peninsula (Blakely and others, 1999b; Blakely and Lowe, 2001) where it can clearly be correlated with Eocene basaltic rocks of the Crescent Formation. Thus, the Devils Mountain fault in this location corresponds to a major crustal boundary between pre-Tertiary basement rocks and younger Tertiary basement rocks (Johnson, 1984b; Johnson and others, 1996).

Evidence for Faulting Onshore, Whidbey and Camano Islands

The approximate locations of the Devils Mountain, Strawberry Point, and Utsalady Point faults on Whidbey Island are inferred based on projections from offshore seismic-reflection data (figs. 8–26), aeromagnetic data (fig. 3), and onshore geologic mapping in the Skagit delta (Whetten and others, 1988). With this information, we examined exposures (fig. 27) and subsurface data (fig. 28, table 1) to more precisely locate faults and constrain fault histories, lengths, and slip rates. Additionally, virtually all coastal-bluff exposures on Whidbey Island and on the northeastern Olympic Peninsula have been examined for this and previous investigations (Johnson and others, 1996, among others) to provide a framework for recognizing structural deformation in Quaternary strata.

Subsurface information comes from lithologic logs of water wells (fig. 28; table 1). Ground water is the primary source of water used on Whidbey and Camano Islands, and ground-water resources have been studied extensively (for example, Anderson, 1968; Cline and others, 1982; Jones, 1985; Sapik and others, 1988), with a focus on aquifer properties, subsurface stratigraphy, and saltwater intrusion. For this study, we used a data base, maintained by the Island County Health Department, that includes logs for more than 3,500 water wells. Stratigraphic interpretation of these logs can be problematic because (1) most wells are quite shallow and were logged by non-geologists, and (2) outcrops of Quaternary strata

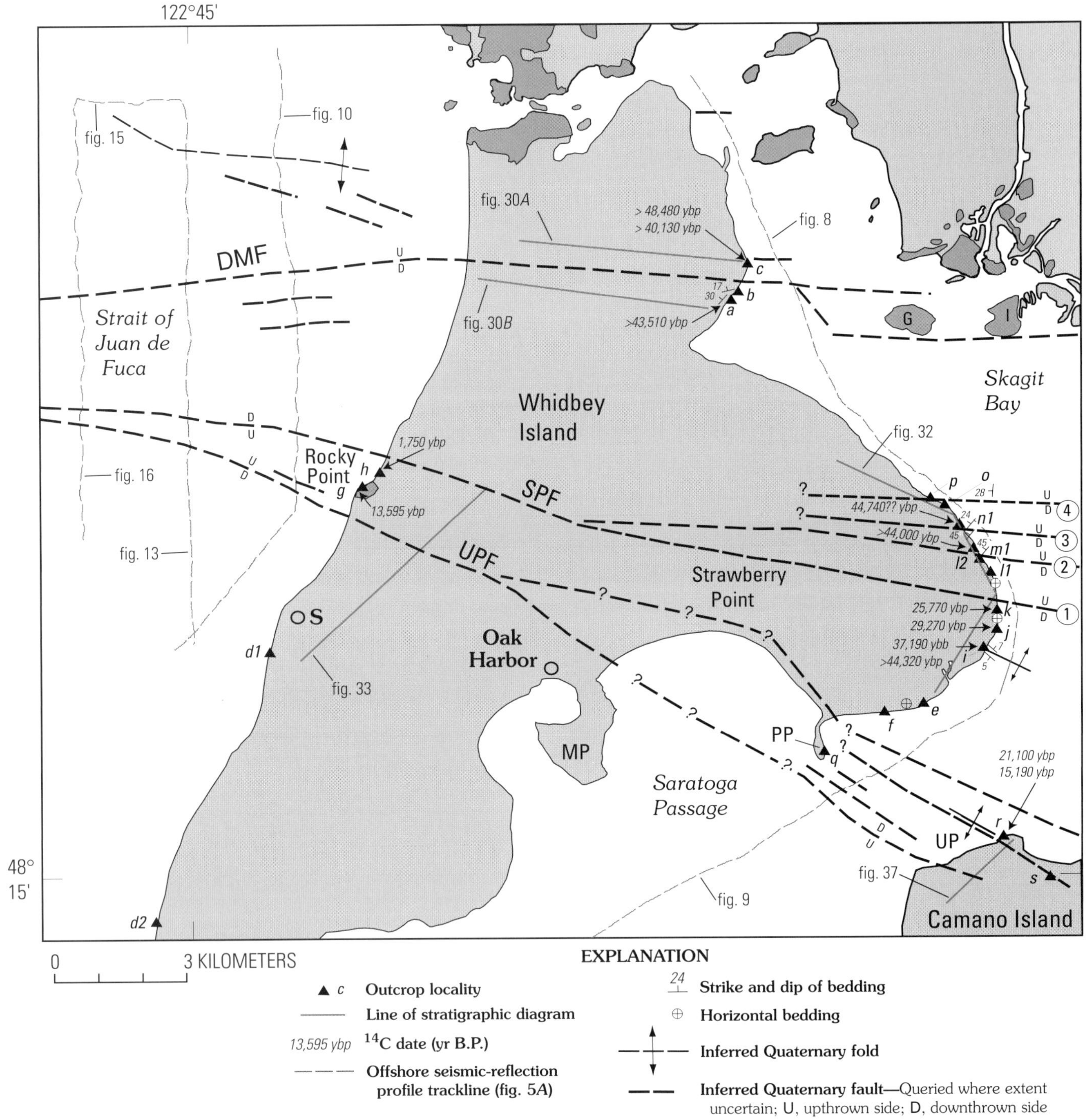

Figure 27. Northern Whidbey Island area showing locations of key outcrop localities described in text (designated by letters and triangles), radiocarbon dates (table 2), nearby seismic-reflection profiles (pl. 1), and faults and folds that deform Quaternary deposits. All dates are described in table 2 except the one obtained at *g*, which is described in Pessl and others (1989). Stratigraphic data from many outcrops were used in constructing stratigraphic correlation diagrams (figs. 30, 32, 33, 37). Circled numbers refer to faults that cut Strawberry Point, as described in text. Dark-shaded areas are underlain by pre-Quaternary bedrock; light-shaded areas are underlain by Quaternary deposits. DMF, Devils Mountain fault; G, Goat Island; I, Ika Island; MP, Maylor Point; PP, Polnell Point; S, Swantown; SPF, Strawberry Point fault; UP, Utsalady Point; UPF, Utsalady Point fault.

throughout the Puget Lowland reveal that substantial erosional relief can occur within and at the upper and lower contacts of glacially influenced depositional units (fig. 6). Despite these potential problems, the water-well data base can have great value in stratigraphic investigations when (1) the lithologic logs are carefully selected based on quality of logging, (2) a large proportion of deeper wells (>50 m) which penetrate several stratigraphic units are used; (3) the number of wells used for correlation on stratigraphic diagrams is large enough to mitigate problems associated with variable erosional relief and with poor logging in a few wells, and (4) information from boreholes can be correlated with nearby outcrops (fig. 27). Table 1 is a list of wells (including depth and location) used to draw stratigraphic correlation diagrams for this investigation.

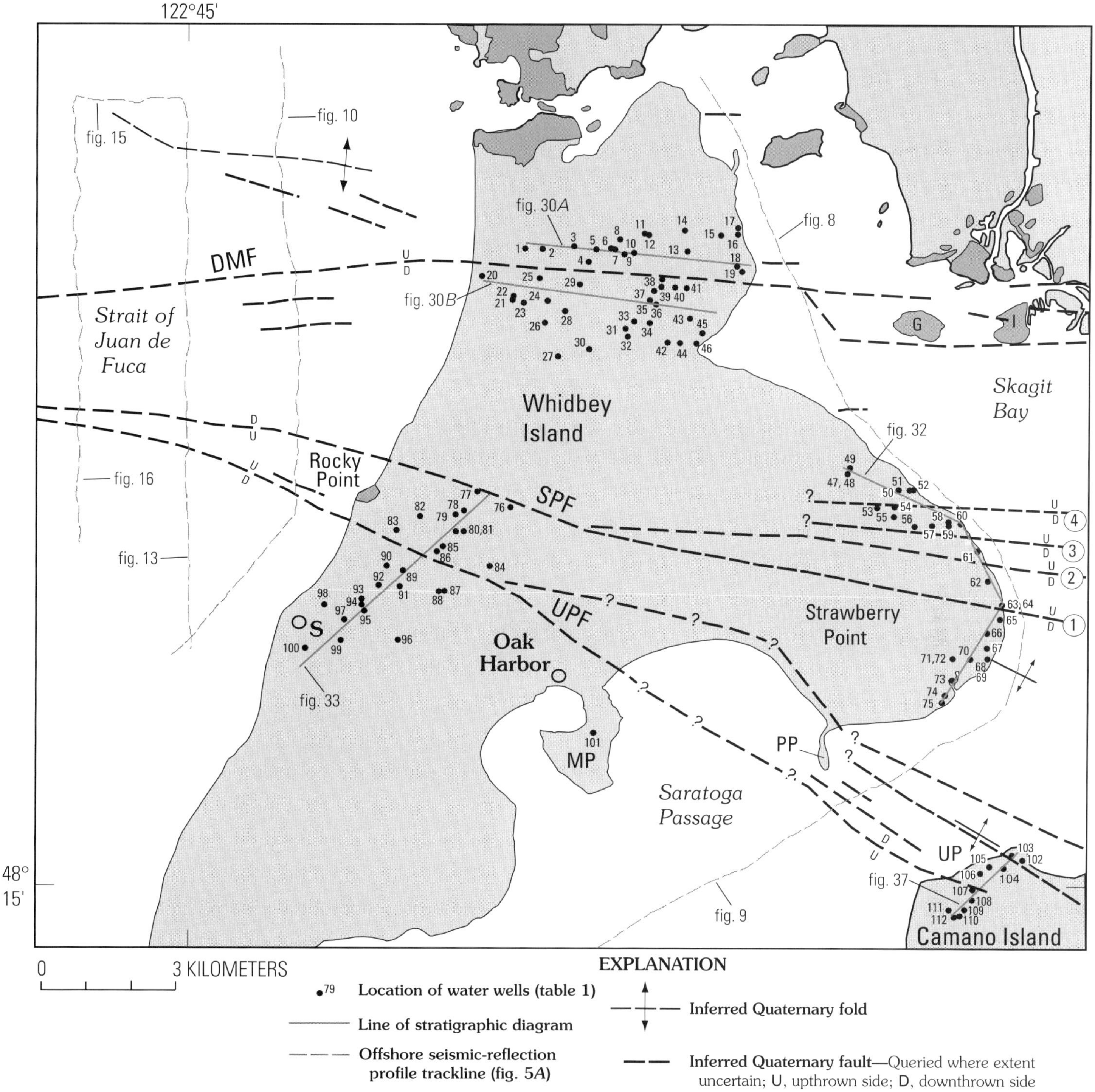

Figure 28. Northern Whidbey Island area showing locations of water wells (designated by numbers and dots) used in making stratigraphic correlation diagrams (table 1; figs. 30, 32, 33, 37) and faults and folds that deform Quaternary deposits. Circled numbers refer to faults that cut Strawberry Point, as described in text. Dark-shaded areas are underlain by pre-Quaternary bedrock; light-shaded areas are underlain by Quaternary deposits. DMF, Devils Mountain fault; G, Goat Island; I, Ika Island; MP, Maylor Point; PP, Polnell Point; S, Swantown; SPF, Strawberry Point fault; UP, Utsalady Point; UPF, Utsalady Point fault.

Devils Mountain Fault

Location

The Devils Mountain fault zone is not exposed on Whidbey Island. Based on projection of seismic-reflection data (figs. 8, 10), the southern fault of the zone (inferred to be the most active strand and the northern margin of the Everett basin) extends through an area characterized by large landslides and variable exposure on Whidbey Island's east coast and a marshy lowland on the island's west coast. Inland, the projected "basin-margin" fault strand is covered by vegetation and surficial deposits developed mainly on Vashon Drift (Pessl and others, 1989). No obvious topographic lineament is associated with the fault where it

Table 1. Information on water wells used in constructing stratigraphic correlation diagrams (figs. 30, 32, 33, 37).

[Well locations are shown in figure 28. Location column shows township/range, section, quarter section. Latitude and longitude are from digital data base maintained by the Island County Health Department. For wells not included in this data base, the location within 1/16 of the section is designated (Anderson, 1968)]

Figure 30A:

Well #	Location	Lat(°)	Long(°)	Depth in feet (m)
1	T33N/R1E, sec. 2, NW 1/4	48.379	122.649	195 (59.4)
2	T33N/R1E, sec. 2, NE 1/4	48.378	122.645	286 (87.2)
3	T33N/R1E, sec. 1, NW 1/4	48.379	122.634	165 (50.3)
4	T33N/R1E, sec. 1, SW 1/4	48.376	122.630	152 (46.3)
5	T33N/R1E, sec. 1, NW 1/4	48.379	122.628	205 (62.5)
6	T33N/R1E, sec. 1, NE 1/4	48.379	122.623	196 (59.7)
7	T33N/R1E, sec. 1, NE 1/4	48.379	122.622	196 (59.7)
8	T33N/R1E, sec. 1, NE 1/4	48.380	122.620	205 (62.5)
9	T33N/R1E, sec. 1, NE 1/4	48.378	122.617	238 (72.5)
10	T33N/R1E, sec. 1, NE 1/4	48.378	122.617	223 (68.0)
11	T33N/R2E, sec. 6, NW 1/4	48.381	122.613	272 (83.0)
12	T33N/R2E, sec. 5, NW 1/4	48.382	122.611	287 (87.5)
13	T33N/R2E, sec. 6, NE 1/4	48.378	122.600	254 (77.4)
14	T33N/R2E, sec. 6, NE 1/4	48.382	122.600	200 (61.0)
15	T33N/R2E, sec. 5, NW 1/4	48.382	122.589	233 (71.0)
16	T33N/R2E, sec. 5, NW 1/4	48.382	122.584	189 (57.6)
17	T33N/R2E, sec. 5, SW 1/4	48.383	122.585	160 (48.8)
18	T33N/R2E, sec. 5, SE 1/4	48.375	122.583	284 (86.6)
19	T33N/R2E, sec. 5, SW 1/4	48.374	122.581	282 (86.0)

Figure 30B:

Well #	Location	Lat(°)	Long(°)	Depth in feet (m)
20	T33N/R1E, sec. 3, SE 1/4	48.377	122.664	109 (33.2)
21	T33N/R1E, sec. 11, NW 1/4	48.369	122.654	95 (29.0)
22	T33N/R1E, sec. 11, NW 1/4	48.369	122.654	158 (48.2)
23	T33N/R1E, sec. 11, NW 1/4	48.368	122.650	152 (46.3)
24	T33N/R1E, sec. 2, SE 1/4	48.368	122.642	243 (74.1)
25	T33N/R1E, sec. 2, SE 1/4	48.373	122.646	214 (65.2)
26	T33N/R1E, sec. 11, NE 1/4	48.364	122.643	213 (64.9)
27	T33N/R1E, sec. 11, SW 1/4	48.357	122.639	92 (28.0)
28	T33N/R1E, sec. 11, NE 1/4	48.366	122.637	246 (75.0)
29	T33N/R1E, sec. 1, SW 1/4	48.371	122.633	140 (42.7)
30	T33N/R1E, sec. 12, SW 1/4	48.359	122.630	48 (14.6)
31	T33N/R1E, sec. 12, NE 1/4	48.362	122.619	155 (47.2)
32	T33N/R1E, sec. 12, SE 1/4	48.361	122.618	118 (36.0)
33	T33N/R1E, sec. 12, NE 1/4	48.364	122.615	108 (32.9)
34	T33N/R2E, sec. 7, NW 1/4	48.364	122.611	139 (42.4)
35	T33N/R2E, sec. 7D, NW 1/4			182 (55.5)
36	T33N/R2E, sec. 6, SW 1/4	48.367	122.611	216 (65.8)
37	T33N/R2E, sec. 6, NE 1/4	48.371	122.609	261 (79.6)
38	T33N/R2E, sec. 6, SE 1/4	48.374	122.607	238 (72.5)
39	T33N/R2E, sec. 6, SW 1/4	48.371	122.608	267 (81.4)
40	T33N/R2E, sec. 6, SW 1/4	48.371	122.604	274 (83.5)
41	T33N/R2E, sec. 6, SW 1/4	48.371	122.600	236 (71.9)
42	T33N/R2E, sec. 7, SW 1/4	48.360	122.606	177 (54.0)
43	T33N/R2E, sec. 7, NE 1/4	48.365	122.598	249 (75.9)
44	T33N/R2E, sec. 7, SW 1/4	48.360	122.602	76 (23.1)
45	T33N/R2E, sec. 7, SE 1/4	48.362	122.595	39 (11.9)
46	T33N/R2E, sec. 7, SE 1/4	48.360	122.597	180 (54.9)

crosses Whidbey Island, in contrast to the conspicuous trace of the fault in the Cascade Range foothills to the east (fig. 4).

The projection of the southern "basin-margin" fault in the Devils Mountain fault zone across Whidbey Island is on strike with a strand of the Devils Mountain fault mapped in the Skagit delta to the east, north of Goat and Ika Islands in Skagit Bay (figs. 2, 27; Whetten and others, 1988). Goat and Ika Islands are underlain by bedrock, however, and clearly lie north of the northern margin of the Everett basin. Moreover, an aeromagnetic anomaly that defines the basin margin (fig. 3) in Skagit Bay lies just south of these islands, on the southern margin of a small, prominent, shallow (<350 m), high-amplitude, northwest-trending aeromagnetic anomaly interpreted as mafic pre-Tertiary rock. Thus, the "basin-margin" fault appears to step south about 1,000 m in Skagit Bay. This step probably coincides with the steep gradient on the northeast margin of the high-amplitude

Table 1—Continued. Information on water wells used in constructing stratigraphic correlation diagrams (figs. 30, 32, 33, 37).

[Well locations are shown in figure 28. Location column shows township/range, section, quarter section. Latitude and longitude are from digital data base maintained by the Island County Health Department. For wells not included in this data base, the location within 1/16 of the section is designated (Anderson, 1968)]

Figure 32:

Well #	Location	Lat(°)	Long(°)	Depth in feet (m)
47	T33N/R2E, sec. 21, SE 1/4	48.331	122.552	198 (60.4)
48	T33N/R2E, sec. 21, SE 1/4	48.331	122.552	221 (67.4)
49	T33N/R2E, sec. 22, SW 1/4	48.332	122.547	175 (53.3)
50	T33N/R2E, sec. 22, SE 1/4	48.327	122.535	180 (54.9)
51	T33N/R2E, sec. 22, SE 1/4	48.327	122.530	157 (47.9)
52	T33N/R2E, sec. 22, SE 1/4	48.324	122.528	145 (44.2)
53	T33N/R2E, sec. 27, NW 1/4	48.324	122.541	208 (63.4)
54	T33N/R2E, sec. 27, NE 1/4	48.324	122.536	176 (53.6)
55	T33N/R2E, sec. 27, NE 1/4	48.322	122.535	183 (55.8)
56	T33N/R2E, sec. 27, NE 1/4	48.320	122.530	182 (55.5)
57	T33N/R2E, sec. 26, SE 1/4	48.320	122.525	262 (79.9)
58	T33N/R2E, sec. 26, NW 1/4	48.321	122.519	419 (127.7)
59	T33N/R2E, sec. 26, NW 1/4	48.320	122.519	149 (45.4)
60	T33N/R2E, sec. 26, NE 1/4	48.320	122.515	261 (79.6)
61	T33N/R2E, sec. 26, SE 1/4	48.315	122.510	160 (48.8)
62	T33N/R2E, sec. 35, NE 1/4	48.310	122.508	141 (43.0)
63	T33N/R2E, sec. 36, NW 1/4	48.306	122.504	420 (128.0)
64	T33N/R2E, sec. 36, NW 1/4	48.306	122.504	260 (79.3)
65	T33N/R2E, sec. 36, SW 1/4	48.303	122.505	300 (91.4)
66	T33N/R2E, sec. 35, SE 1/4	48.299	122.510	180 (54.9)
67	T33N/R2E, sec. 35, SE 1/4	48.297	122.510	201 (61.3)
68	T32N/R2E, sec. 2, NE 1/4	48.295	122.510	256 (78.0)
69	T32N/R2E, sec. 2, NE 1/4	48.295	122.510	191 (58.2)
70	T32N/R2E, sec. 2, NE 1/4	48.295	122.510	185 (56.4)
71	T32N/R2E, sec. 2, NW 1/4	48.295	122.520	196 (59.7)
72	T32N/R2E, sec. 2, NW 1/4	48.295	122.520	138 (42.1)
73	T32N/R2E, sec. 2, NW 1/4	48.291	122.520	300 (91.4)
74	T32N/R2E, sec. 2, SW 1/4	48.288	122.520	190 (57.9)
75	T32N/R2E, sec. 2, SW 1/4	48.285	122.560	163 (49.7)

Figure 33:

Well #	Location	Lat(°)	Long(°)	Depth in feet (m)
76	T33N/R1E, sec. 26, NW 1/4	48.326	122.650	682 (207.9)
77	T33N/R1E, sec. 22, SE 1/4	48.329	122.660	534 (162.8)
78	T33N/R1E, sec. 27, NE 1/4	48.325	122.670	159 (48.5)
79	T33N/R1E, sec. 27, NW 1/4	48.324	122.670	200 (61.0)
80	T33N/R1E, sec. 27, NE 1/4	48.321	122.670	250 (76.2)
81	T33N/R1E, sec. 27, NE 1/4	48.321	122.670	315 (96.0)
82	T33N/R1E, sec. 28, NE 1/4	48.324	122.680	366 (111.6)
83	T33N/R1E, sec. 28, NE 1/4	48.321	122.690	261 (79.6)
84	T33N/R1E, sec. 27, SE 1/4	48.314	122.660	208 (63.4)
85	T33N/R1E, sec. 27, SW 1/4	48.318	122.670	269 (82.0)
86	T33N/R1E, sec. 27, SW 1/4	48.317	122.680	240 (73.2)
87	T33N/R1E, sec. 34, NW 1/4	48.309	122.673	183 (55.8)
88	T33N/R1E, sec. 34, NW 1/4	48.309	122.675	174 (53.0)
89	T33N/R1E, sec. 28, SE 1/4	48.313	122.686	193 (58.8)
90	T33N/R1E, sec. 28L, SW 1/4			194 (59.1)
91	T33N/R1E, sec. 33, NE 1/4	48.310	122.687	134 (40.8)
92	T33N/R1E, sec. 33, NW 1/4	48.310	122.693	162 (49.4)
93	T33N/R1E, sec. 33, NW 1/4	48.307	122.698	286 (87.2)
94	T33N/R1E, sec. 33, NW 1/4	48.306	122.698	287 (87.5)
95	T33N/R1E, sec. 33, SW 1/4	48.305	122.697	320 (97.5)
96	T33N/R1E, sec. 33, SE 1/4	48.299	122.687	245 (74.7)
97	T33N/R1E, sec. 32, SE 1/4	48.303	122.703	280 (85.3)
98	T33N/R1E, sec. 32, NE 1/4	48.306	122.709	375 (114.3)
99	T33N/R1E, sec. 32, SE 1/4	48.299	122.704	249 (75.9)
100	T32N/R1E, sec. 5, NW 1/4	48.298	122.715	1000 (304.8)

Table 1—Continued. Information on water wells used in constructing stratigraphic correlation diagrams (figs. 30, 32, 33, 37).

[Well locations are shown in figure 28. Location column shows township/range, section, quarter section. Latitude and longitude are from digital data base maintained by the Island County Health Department. For wells not included in this data base, the location within 1/16 of the section is designated (Anderson, 1968)]

Deep test well:

Well #	Location	Lat(°)	Long(°)	Depth in feet (m)
101	T32N/R1E, sec. 12, NW 1/4	48.280	122.629	1933 (589.2)

Figure 37:

Well #	Location	Lat(°)	Long(°)	Depth in feet (m)
102	T32N/R2E, sec. 24, NW 1/4	48.254	122.500	213 (64.9)
103	T32N/R2E, sec. 14, SE 1/4	48.255	122.503	187 (57.0)
104	T32N/R2E, sec. 23, NE 1/4	48.253	122.506	150 (45.7)
105	T32N/R2E, sec. 23, NE 1/4	48.253	122.510	280 (85.3)
106	T32N/R2E, sec. 23, NE 1/4	48.252	122.512	400 (121.9)
107	T32N/R2E, sec. 23, NW 1/4	48.249	122.515	320 (97.5)
108	T32N/R2E, sec. 23, SW 1/4	48.247	122.515	340 (103.6)
109	T32N/R2E, sec. 23, SW 1/4	48.246	122.517	370 (112.8)
110	T32N/R2E, sec. 23, SW 1/4	48.245	122.516	182 (55.5)
111	T32N/R2E, sec. 23, SW 1/4	48.246	122.519	158 (48.2)
112	T32N/R2E, sec. 23, SW 1/4	48.244	122.518	145 (44.2)

aeromagnetic anomaly, and may represent a transfer zone or tear fault in the fault system. In this scenario, the inferred mafic rock is a basement block uplifted within this transfer zone.

Outcrops

Pleistocene strata are exposed near the fault along the east coast of Whidbey Island. Locality *a* (fig. 27), south of the fault trace, consists of three outcrops exposed within 100 m in coastal bluffs or on the beach at low tide. Beds consisting of sand, organic-rich mud, and gravel strike N. 25°–35° E. and dip 25°–36° NW. Detrital wood yielded an infinite radiocarbon date (>43,510 ^{14}C yr B.P.; SJ-00-7; table 2). About 200 m farther north (*b* in fig. 27), similar sand and gravel facies exposed in a landslide scarp strike 70° and dip 17° NW.

About 300 m north of the inferred fault trace (*c* in fig. 27), Quaternary strata exposed in the upper part of steep coastal bluffs are subhorizontal (fig. 29). The lowest beds at this locality consist of interbedded fine to coarse sand (plane-bedded and crossbedded), laminated silty mud, and peat. Based on their sedimentology, these strata are interpreted as interglacial alluvial-plain deposits. The peat yielded radiocarbon dates of >48,480 ^{14}C yr B.P. (SJ-98-17; table 2) and >40,130 ^{14}C yr B.P. (SJ-00-1; table 2). These strata, and the peat-bearing strata exposed at *a*, are correlated with the Whidbey Formation, the oldest known interglacial alluvial-plain deposits recognized on Whidbey Island and in the northern Puget Lowland (Easterbrook, 1994a, b). The correlation is based on three factors:

1. The lithology and sedimentology of these strata are identical to those of well-documented Whidbey Formation localities elsewhere on Whidbey Island. (See, for example, Easterbrook, 1968, 1994b; Stoffel, 1981.)

2. Subsurface data (fig. 30, see next section) from this local northern Whidbey Island area show that strata occurring at similar elevations and with similar lithology are overlain by two

Figure 29. Bluff outcrops of Quaternary strata at locality *c* (fig. 27). Annotation shows units (fig. 6) and contacts. O, Olympia beds; P, Possession Drift; W, Whidbey Formation. Two infinite radiocarbon dates from organic matter in the Whidbey Formation were obtained at this locality (table 2, fig. 27).

Table 2. Radiocarbon data for samples from northern Whidbey and Camano Islands.
[Laboratory data are from Beta Analytic Inc.]

Sample Number	Radiocarbon laboratory Number	Field locality (fig. 27)	Radiocarbon age (^{14}C yr B.P.)	$^{13}C/^{12}C$ (o/oo)	Calibrated age (calendar year, 95.4% confidence)	Material dated	Comment
SJ-98-17	120288	*c*	>48,480	-25.6	NA[1]	Peat	Whidbey Formation north of Devils Mountain fault.
SJ-98-14	120287	*h*	1,750±50	-25.8	A.D. 210-420	Wood in peat	At Rocky Point between Strawberry Point and Utsalady Point faults.
SJ-99-2	134854	*m*	>44,000	-27.9	NA[1]	Peat	Whidbey Formation in Strawberry Point fault zone at Strawberry Point.
SJ-99-1A	132512	*r*	21,100±150	-24.9	NA[1]	Organic laminae	Nonglacial Olympia beds at Utsalady Point.
SJ-99-1B	141690	*r*	15,190±220	-23.6	17,435-18,985 B.P.	Organic laminae	Nonglacial Olympia beds at Utsalady Point.
SJ-00-1	144555	*c*	>40,130	-28.2	NA[1]	Peat	Whidbey Formation north of Devils Mountain fault.
SJ-00-7	144556	*a*	>43,510	-26.9	NA[1]	Wood in peat	Whidbey Formation south of Devils Mountain fault.
SJ-00-11	144557	*i*	>44,320	-25.6	NA[1]	Peat	Whidbey Formation south of Strawberry Point fault zone at Strawberry Point.
SJ-00-12	144558	*j*	29,270±340	-27.6	NA[1]	Peat	Olympia beds south of Strawberry Point fault zone at Strawberry Point.
SJ-00-13	144559	*k*	25,770±310	-28.4	NA[1]	Peat	Olympia beds south of Strawberry Point fault zone at Strawberry Point.
SJ-00-14	144554	*n*	44,740±910	-28.9	NA[1]	Detrital wood	Whidbey Formation in Strawberry Point fault zone at Strawberry Point; young age suggests probable contamination.
SJ-00-48	147710	*i*	37,190±550	-25.8	NA[1]	Detrital wood	Basal Olympia beds south of Strawberry Point fault zone at Strawberry Point.

[1]NA, not applicable.

units of interbedded "hardpan" (the term drillers use for dense clay-rich diamict), gravel, and sand interpreted as glacial drift. The Whidbey Formation predates both the Possession and Vashon Drifts, whereas the younger nonglacial Olympia beds predate only the Vashon Drift (fig. 6). At *c*, the Whidbey Formation is overlain by crudely stratified gravel assigned to the Possession Drift and interbedded sand (parallel bedded and crossbedded) and silt assigned to the nonglacial Olympia beds (fig. 6).

3. Peat from these strata is too old (> 48,480 ^{14}C yr B.P.) to yield reliable radiocarbon ages (table 2), consistent with previous dating results from the Whidbey Formation (Blunt and others, 1987). In addition to the dates reported here, J.D. Dragovich (written commun., 2000) obtained four radiocarbon dates from the same unit at four localities within 2.5 km north of location *c* (fig. 27) along Whidbey Island's northeast coast. Two dates from wood samples similarly yielded infinite ages (>43,790 ^{14}C and >37,100 ^{14}C yr B.P.), whereas an organic silty sand sample yielded a finite age of 41,380±2,150 ^{14}C yr B.P. and organic material in silt yielded a finite age of 40,590±530 ^{14}C yr B.P. These two finite ages from organic sediment are very close to the practical limit of radiocarbon dating, and they may have been contaminated by minute amounts of younger carbon.

Subsurface Data

East-west stratigraphic correlation diagrams on the north and south sides of the inferred trace of the Devils Mountain fault on northern Whidbey Island (locations, figs. 27, 28) are shown in figure 30. The diagram north of the fault includes 19 wells within 1.4 km of the inferred fault trace and one surface outcrop (figs. 29, 30*A*). Stratigraphic units inferred on the correlation diagram include pre-Whidbey Formation Pleistocene deposits, the Whidbey Formation, Possession Drift, Olympia beds, Esperance Sand Member, main body of Vashon Drift, and Everson Drift (fig. 6). Lateral stratigraphic variation across this diagram probably reflects irregular scour and fill associated with the Possession and Vashon glaciations, and possible faulting within the block north of the Devils Mountain fault (fig. 8*A*), but it may also reflect incorrect stratigraphic correlation. Interpretations on the correlation diagram south of the Devils Mountain fault (fig. 30*B*), which includes data from 27 wells and one outcrop proximal to the inferred fault trace, suggest greater lateral stratigraphic continuity.

The stratigraphic correlation diagrams (fig. 30*A*, *B*) were generated to provide constraints on vertical offset across the Devils Mountain fault. In this regard, lateral variability and potentially uncertain correlation of units above the Whidbey Formation make their utility as deformation markers problematic. The top of the interglacial Whidbey Formation is inferred to provide the most useful datum for the following reasons:

1. The top of the Whidbey Formation is a distinct contact in both outcrops (for example, Easterbrook, 1968) and subsurface data (fig. 30), generally characterized by an abrupt upward transition from interbedded sand, silt, clay, and peat to overlying interbedded gravel, sand, and "hardpan."

2. The Whidbey Formation is relatively thick and widespread on Whidbey Island (Easterbrook, 1968), and is present in every well that reaches its stratigraphic level. This occurrence reflects its deposition during a period of relatively high sea level (marine isotope stage 5; Muhs and others, 1994; Pillans and others, 1998), a condition which generates significant stratigraphic "accommodation space" (Jervey, 1988; Posamentier and others, 1988) in which sediments can accumulate and have high potential for preservation in the stratigraphic record. In contrast, younger (marine isotope stage 3) nonglacial Olympia beds (fig. 6) are absent (due to nondeposition or erosion) in several wells and in most outcrops on Whidbey Island. Sea level at this time was much lower than at present and in stage 5 (Pillans and others, 1998), significantly limiting both stratigraphic accommodation space and preservation potential.

3. As an alluvial-plain deposit, the top of the Whidbey Formation was a nearly horizontal surface across northern Whidbey Island following deposition. At the end of Whidbey deposition, paleo-relief on this surface in local areas like this small part of northern Whidbey Island would have been at most a few meters, reflecting erosion and deposition associated with fluvial channels and levees. Modern alluvial-delta plains of the Puget Lowland (such as the Skagit delta a few kilometers east of Whidbey Island) provide a useful analog.

4. In both correlation diagrams, the top of the Whidbey Formation is interpreted as a gently undulating though relatively horizontal surface; thus erosion into the Whidbey Formation by advance outwash facies of the Possession Drift appears to have been fairly limited and (or) uniform across this small area of northern Whidbey Island. Widespread preservation of this relatively horizontal surface in areas not within fault zones on northern Whidbey Island is documented by two stratigraphic diagrams in Easterbrook (1968; his fig. 15 and plate 1-*B*) from the coastal bluffs south of Swantown (*d1* and *d2* in fig. 27). These correlation diagrams show the top of the Whidbey Formation as a nearly horizontal surface for 4.6 km (elevation of 17 m) and 3.5 km (elevation of 10 m) of continuous exposure. However, elsewhere (such as *f* in fig. 27), there is locally as much as 20 m of erosional relief on the top of the Whidbey Formation.

On the northern correlation diagram (fig. 30*A*), our interpretation is that the top of the Whidbey Formation was penetrated by 14 wells for which it has a maximum elevation of 38.4 m and a mean elevation of 26.6±5.8 m. On the southern correlation diagram (fig. 30*B*), our interpretation is that the top of the Whidbey Formation was penetrated by 16 wells for which it has a maximum elevation of 26.9 m and a mean elevation of 14.7±6.0 m. The elevation south of the Devils Mountain fault is similar to that observed in outcrops south of Swantown (*d1* in fig. 27; Easterbrook, 1968), whereas this contact is much higher north of the Devils Mountain fault. The difference between the maximum and mean elevations on the two correlation diagrams is 11.5 m and 11.9 m, respectively. The mean and standard deviation of the sum and difference of the mean depths (Spiegel, 1961) is 11.9 m±8.3 m. If one assumes that the amount of erosional relief on this datum is similar north and south of the projected fault trace, then these values provide a rough estimate for the amount of vertical separation along the Devils Mountain fault since deposition of the top of the Whidbey Formation.

Strawberry Point Fault

West Coast of Whidbey Island

There are no exposures where the trace of the Strawberry Point fault projects onto the west coast of Whidbey Island, about 1 km north of Rocky Point (figs. 27, 28). North of the projected trace, no exposures exist for several kilometers. Within about 1 km south of the inferred fault trace, Pleistocene glacial deposits occur in discontinuous bluff exposures, and pre-Tertiary bedrock and Holocene peat crop out at beach level at Rocky Point (*g* in fig. 27). The bedrock consists of Upper Cretaceous graywacke, argillite, greenstone, and chert (Whetten and others, 1988) and is extensively faulted and foliated. These rocks have surprisingly low bulk magnetic susceptibility—0.53×10^{-3} (SIU) based on 30 in-place measurements—and form a distinct and continuous west-northwest-trending aeromagnetic low over Rocky Point (fig. 3). In contrast, Quaternary deposits that crop out adjacent to bedrock exposures yield susceptibilities that are roughly an order of magnitude higher.

Pleistocene strata (2 to 15–20 m thick) that overlie the bedrock at *g* (fig. 27) consist of horizontal to very gently dipping (0°–2°) Everson Drift (fig. 6) that yielded a ^{14}C date on shells of 13,595±145 ^{14}C yr B.P. (Pessl and others, 1989). Older Pleistocene strata, more than 450 m thick elsewhere on northern Whidbey Island (as in well 101, fig. 28, table 1), are absent at this locality.

A woody peat is exposed south of the inferred fault trace at locality *h* (fig. 27) for about 200 m along the beach at very low tides (≤40 cm below mean low low water). A fragment of tree branch (2 cm diameter) from this peat yielded a ^{14}C date of 1,750+/-50 ^{14}C yr B.P. (SJ-98-14; table 2). This peat includes rhizomes of either seacoast bulrush (*Scirpus maritimus*) or hardstem bulrush (*Scirpus acutus*), both of which generally occur in the middle and upper parts of brackish or salt marshes (elevation ≥tide of 100 cm) (Cooke, 1997). The presence of these rhizomes is consistent with a relative sea-level rise of about 1.5 to 3 m since about 1.7 ka at this locality. This rise supports regional investigations (for example, Clague and others, 1982; Eronen and others, 1987) that suggest rates of sea-level rise of 1 mm/yr or less for the last few thousand years in Puget Sound and southern British Columbia.

There are an insufficient number of deep water wells flanking the Strawberry Point fault in this area to allow construction of well-constrained stratigraphic correlation diagrams either parallel to (as in fig. 30) or across the fault in order to constrain fault slip.

Central Whidbey Island

Seismic-reflection data (figs. 8, 10, 13) suggest that the Strawberry Point fault changes from a discrete fault into a broad fault zone as it crosses Whidbey Island from west to east (figs. 2, 27, 28). On central Whidbey Island, the trace of this zone is covered by vegetation and surficial deposits developed mainly on Vashon Drift and recessional outwash (Pessl and others, 1989). There are no obvious topographic lineaments associated with the fault zone on existing topographic maps in this area. Furthermore, the number of deep water wells flanking the Strawberry Point fault in this area is insufficient to construct stratigraphic correlation diagrams either parallel to or across the fault in order to constrain fault location or slip. The location of the fault zone and its components on central Whidbey Island is thus somewhat speculative, and is based primarily on aeromagnetic data (fig. 3; Blakely and Lowe, 2001).

East Coast of Whidbey Island

Seismic-reflection data (fig. 8) suggest that the Strawberry Point fault is a zone comprising several splays along Strawberry Point on the east coast of Whidbey Island. Within this zone, there is significant deformation of upper Pleistocene strata discontinuously exposed in coastal bluffs (fig. 31). There are also a significant number of lithologic logs from water wells adjacent to the coastline (table 1, fig. 28). Together, the outcrops, lithologic logs, and seismic-reflection profiles provide the data needed for a stratigraphic correlation diagram across Strawberry Point that provides some constraints on fault-zone interpretation (fig. 32).

Easterbrook (1968), Hansen and Easterbrook (1974), and Blunt and others (1987) have described aspects of the stratigraphy and palynology of the nonglacial Olympia beds exposed at Strawberry Point, and reported dates ranging from 22,700 to 47,600 ^{14}C yr B.P. The unit consists mainly of mud and peat but also includes coarse sand, gravel, and redeposited peat that may have been deposited in Skagit River floodplains and paleochannels, graded to a lower sea level. Younger deposits (Esperance Sand Member and main body of Vashon Drift) locally overlie these strata in the coastal bluffs (fig. 31*A*). The nonglacial Olympia beds are underlain by a variably thick unit of diamict, gravel, and sand assigned to the Possession Drift, in turn underlain by a fine-grained unit consisting of interbedded ripple- and plane-laminated sand, silty mud, and peat. This lower fine-grained unit occurs in the axis of a gentle anticline on the southeast coast of Strawberry Point and in faulted and folded strata on the northeast coast. These older strata are assigned to the Whidbey Formation because (1) radiocarbon dates yield infinite ages (see next paragraph and table 2), (2) they underlie the well-dated Olympia beds and Possession Drift (fig. 6), (3) lithologies are similar to those described for the Whidbey Formation at nearby Whidbey Island localities (for example, Easterbrook, 1968), and (4) no similar nonglacial units older than the Whidbey Formation have been identified in this part of the Puget Lowland. Pebbly diamict that underlies the Whidbey Formation in Strawberry Point water wells is considered pre-Whidbey Formation glacial drift (figs. 6, 32).

Figures 27 and 28 show where the four faults identified on seismic-reflection data (fig. 8) are projected to intersect the coastline. South of fault 1, the exposed Pleistocene section (fig. 31*A*) is folded in a gentle anticline with the Whidbey Formation in its core at sea level, an interpretation supported by well data (fig. 32). Outcrops of Whidbey Formation and Possession Drift on the south limb of the fold dip 5°, and outcrops of Possession Drift and nonglacial Olympia beds on the north limb of the fold (*i* in fig. 27) dip 7°. Peat in the Whidbey Formation exposed at beach level at *i* yielded a radiocarbon date of >44,320 ^{14}C yr B.P. Both low-tide beach exposures and the figure 32 correlation diagram suggest that dips on the north limb of the fold may be steeper than 7° in the Whidbey Formation and older strata, but this interpretation is based on outcrop with minimal stratigraphic context and projection of only one well (No. 65), and is thus tentative. The section exposed at sea level on the north limb of the fold extends upward from the Whidbey Formation through thin Possession Drift into well-dated, nonglacial, peat-bearing Olympia beds (Hansen and Easterbrook, 1974) and is about 30 m thick, providing a minimum estimate for the amount of vertical fold growth since the top of the Whidbey Formation was deposited. For this investigation, we obtained radiocarbon dates of 37,190±550 ^{14}C yr B.P. at *i*, 29,270±340 ^{14}C yr B.P. at *j*, and 25,770±310 ^{14}C yr B.P. at *k* from the Olympia beds (fig. 27). Exposures of the Olympia beds are flat-lying and nearly continuous north of the gentle fold and south of the projected trace of fault 1 (from *j* to *k* in fig. 27).

The projected traces of both faults 1 and 2 are covered by vegetation. However, between these faults, flat-lying massive diamict and crudely stratified gravel and sand crop out extensively (*l1* in fig. 27) in low coastal bluffs, with a maximum exposed thickness of 5–6 m. These strata have an inferred glacial origin based on their texture, and underlie a fine-grained interval in a nearby water well (64, figs. 28, 32; table 1). In the simplest interpretation, the strata in these outcrops are Possession Drift and the overlying fine-grained strata correlate with the Olympia beds. In this scenario (hypothesis 1), vertical offset on fault 1 has raised the basal contact of the Olympia beds about 3–9 m on the north side of the fault. The vertical separation on the top of the older Whidbey Formation based on well logs (figs. 28, 32), also south side down, is about 10–15 m. Alternatively, the coarse-grained section that occurs at sea level between faults 1 and 2 could underlie the Whidbey Formation. In this less likely scenario, south-side-down vertical separation on fault 1 increases to about 40–60 m.

Fault 2 coincides with a significant dip change in coastal outcrops. The last exposure south of the fault (*l2*, ≈20 m south of the fault shown in figs. 27 and 28) consists of flat-lying,

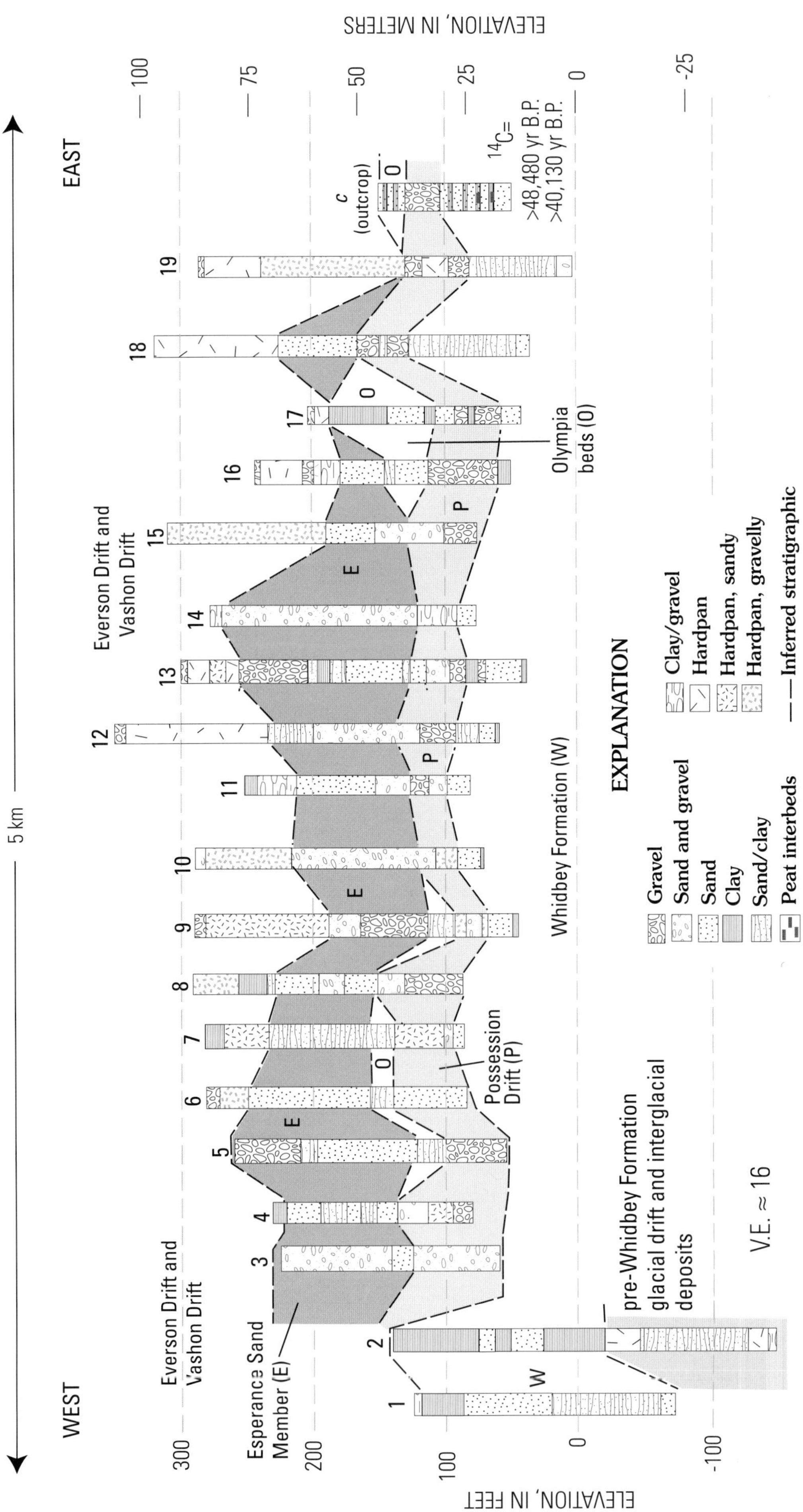
WEST
EAST
5 km
ELEVATION, IN FEET
ELEVATION, IN METERS
300
200
100
0
-100
100
75
50
25
0
-25
1
2
3
4
5
6
7
8
9
10
11
12
13
14
15
16
17
18
19
Everson Drift and Vashon Drift
Esperance Sand Member (E)
pre-Whidbey Formation glacial drift and interglacial deposits
W
E
O
P
Possession Drift (P)
Whidbey Formation (W)
Everson Drift and Vashon Drift
Olympia beds (O)
c (outcrop)
^{14}C=
>48,480 yr B.P.
>40,130 yr B.P.
V.E. ≈ 16
EXPLANATION
Gravel
Sand and gravel
Sand
Clay
Sand/clay
Peat interbeds
Clay/gravel
Hardpan
Hardpan, sandy
Hardpan, gravelly
Inferred stratigraphic contact
A

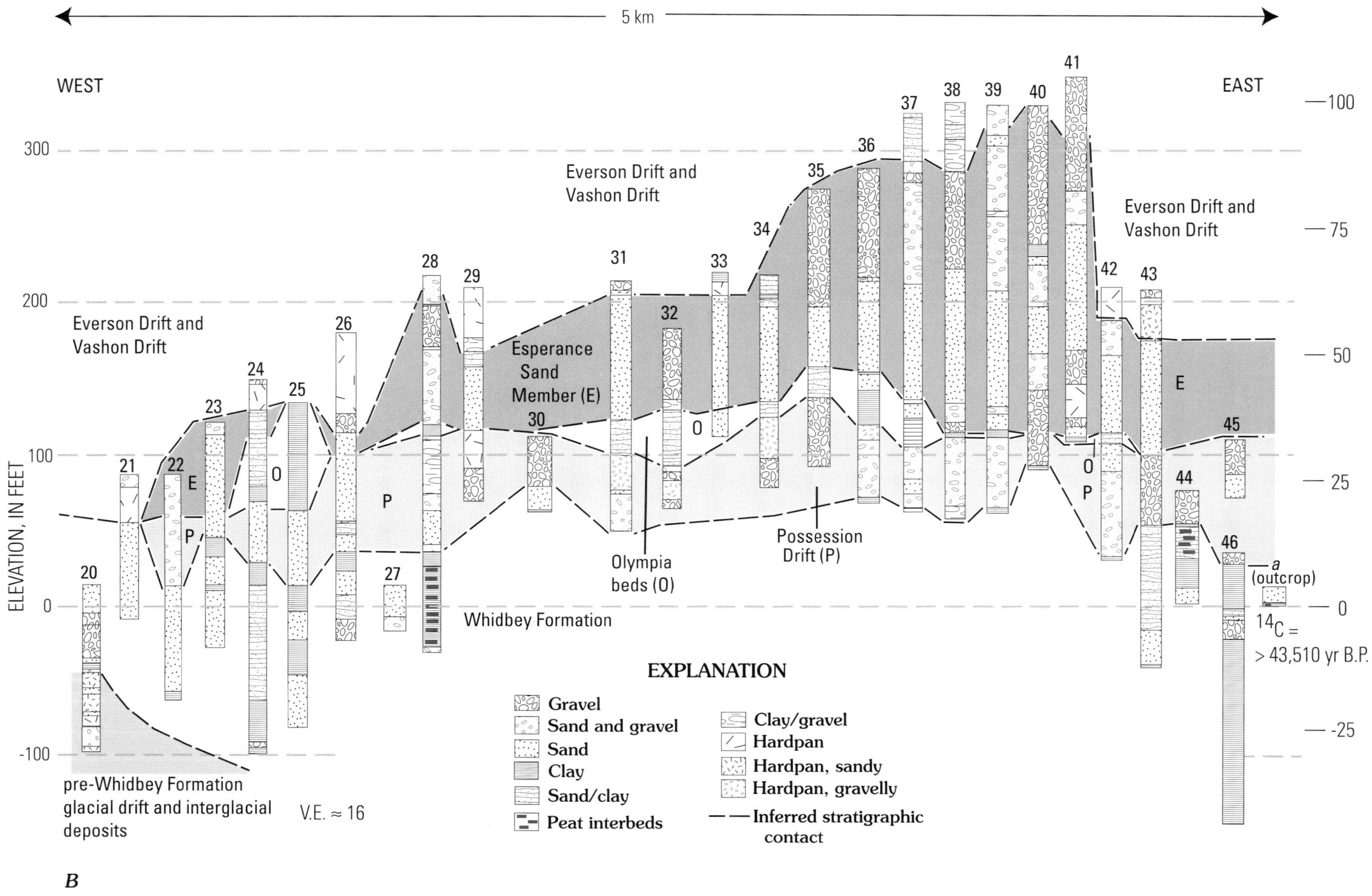

Figure 30. Interpretive east-west stratigraphic correlation diagrams north (*A*) and south (*B*) of the Devils Mountain fault. Information is from outcrops (fig. 27, lettered sections) and lithologic logs from water wells (fig. 28 and table 1, numbered sections), northern Whidbey Island.

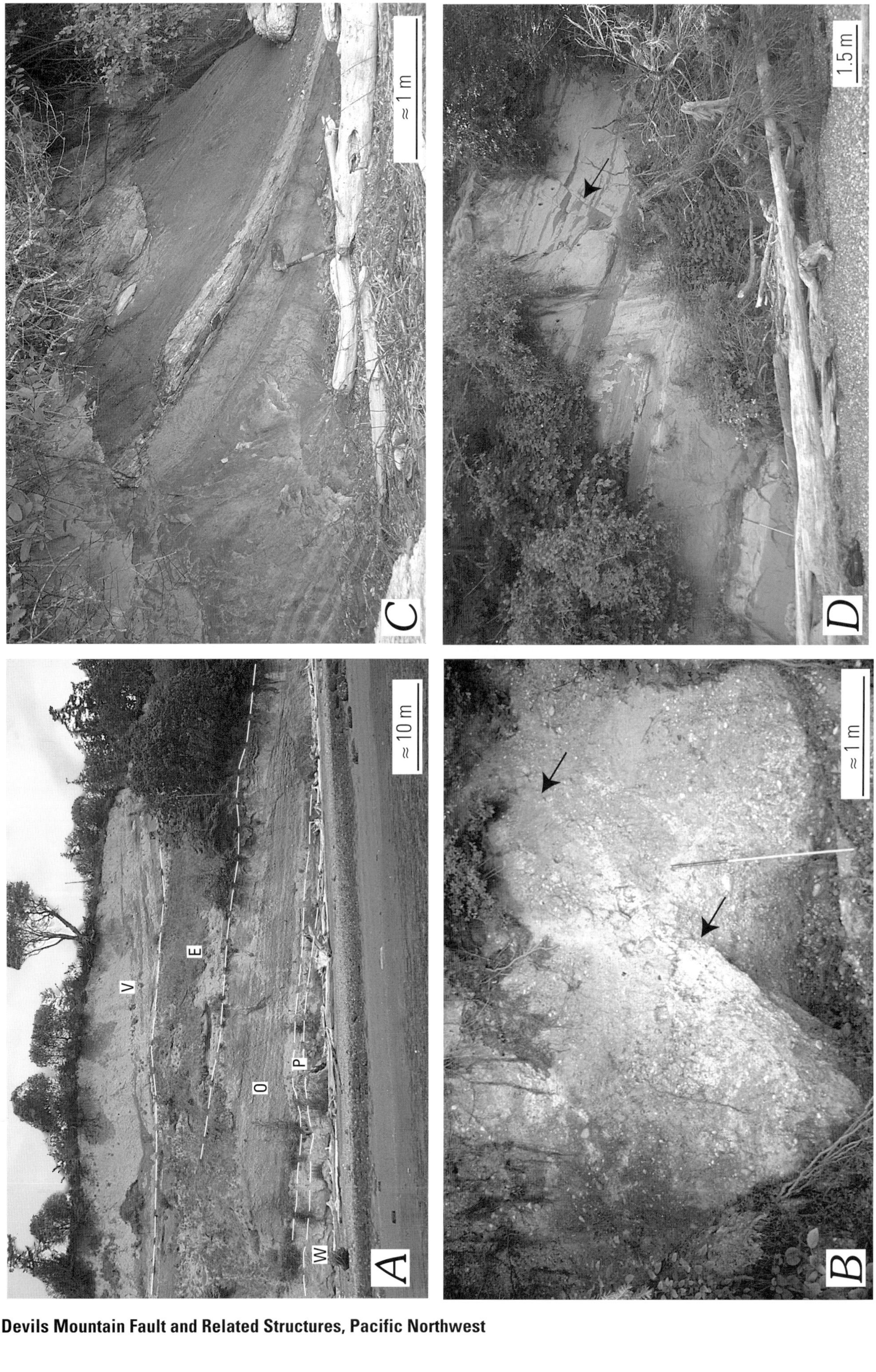
A
V
E
O
P
W
≈10 m
B
≈1 m
C
≈1 m
D
1.5 m

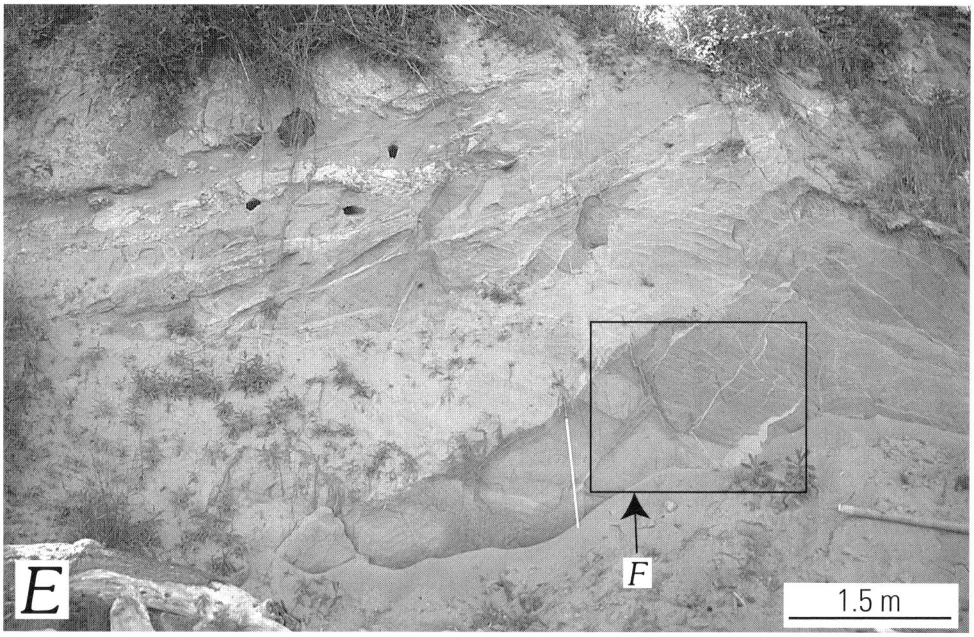

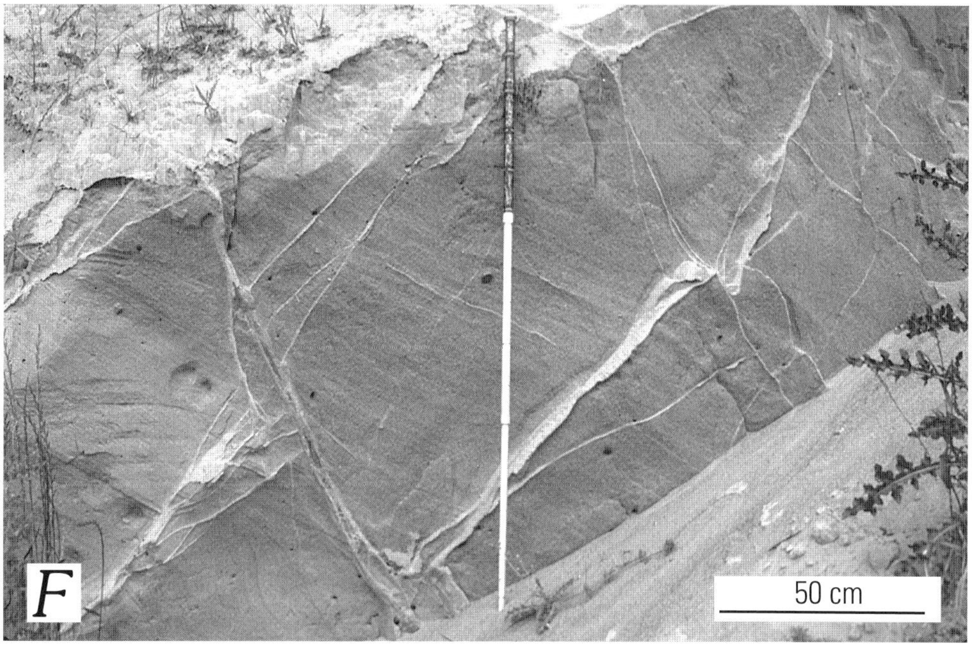

Figure 31 (above and facing page). Deformation of upper Pleistocene strata at Strawberry Point. *A*, Gently dipping (shown by white dashed lines) glacial and interglacial deposits on the north flank of a gentle anticline exposed south of fault 1 at *i*, Strawberry Point (fig. 27). Interbedded sand and mud at the base of the bluff are assigned to the Whidbey Formation (W); organic matter in these beds yielded an infinite radiocarbon date (table 2, fig. 27). These strata are overlain by a thin diamict assigned to the Possession Drift (P), in turn overlain by mud, sand, and gravel assigned to the nonglacial Olympia beds (O). Detrital peat at the base of the Olympia at this locality yielded a date of 37,190± 550 ^{14}C yr B.P. The Olympia is overlain by Esperance Sand Member (E) and main body of Vashon Drift (V). *B*, Gravel and sand assigned to the Possession Drift cut by a thrust fault (arrows) with displacement of 1–2 m, approximately 40 m south of *m* (fig. 27). *C*, Whidbey Formation strata dipping ≈45° NW., locality *m* on Strawberry Point. Beds consist of mud, sand, and woody peat (white-weathering bed above shovel). Peat at this locality yielded an infinite radiocarbon date. Shovel handle is divided into 10 cm increments. *D*, Exposure of Whidbey(?) Formation sand and mud at locality *o* on Strawberry Point (fig. 27). Beds strike north, dip 28° W., and are cut by subvertical fractures (see arrow in upper right). *E*, Exposure of highly fractured friable sand of the Whidbey(?) Formation at locality *p* on Strawberry Point (fig. 27). Box shows approximate location of *F*. *F*, Close-up of outcrop at *p* (see *E*), showing highly fractured friable sand.

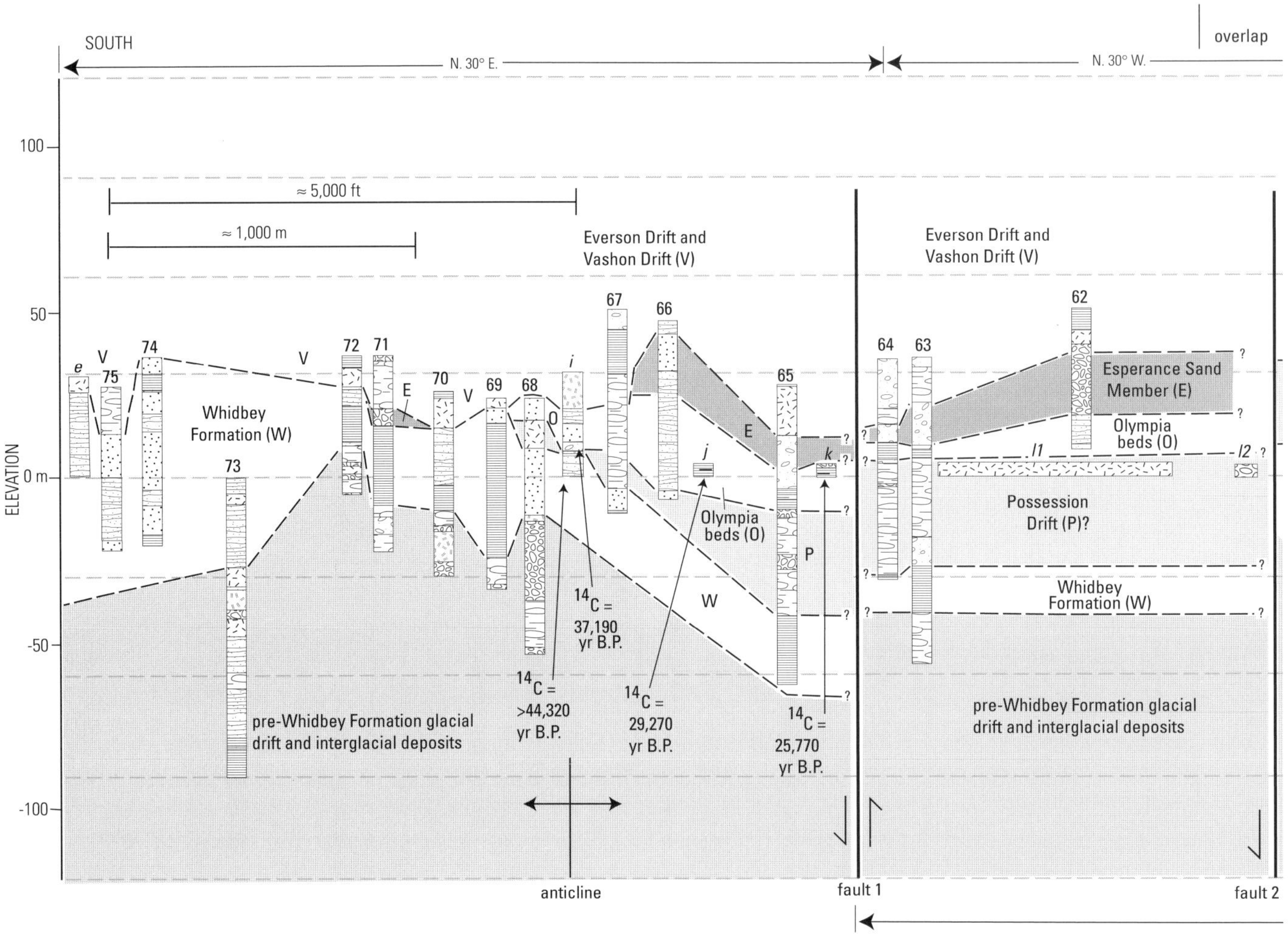

Figure 32. Interpretive stratigraphic correlation diagram parallel to coast of Strawberry Point crossing Strawberry Point fault zone, using outcrop data (figs. 27, 31; lettered sections) and lithologic logs from water wells (table 1, fig. 28; numbered sections), northern Whidbey Island. Faults queried where uncertain; barbs show sense of relative fault offset. Note overlap between the two panels of this figure.

crudely stratified sand and gravel assigned to the Possession Drift. These strata are cut by a thrust fault that strikes N. 77° E., dips 45° NW., and has apparent displacement of 1–2 m (fig. 31*B*).

The first exposures north of fault 2 (*m1* in figs. 27, 32) consist of a continuous 33-m-thick section of sand, silt, and peat assigned to the Whidbey Formation that strike N. 35° E., and dip 42°–48° NW. (fig. 31*C*). Peat from these strata yielded a radiocarbon date of >44,000 ^{14}C yr B.P. (fig. 27, table 2). The section is cut by numerous fractures that generally trend northeast (10°–50°) and dip steeply (>70°) to the southeast. Offsets of 1–15 cm are common on these fractures, and range up to 100 cm. These Whidbey Formation strata are overlain by a 10-m-thick covered interval, then a continuous 60-m-thick section of sand and gravel assigned to the Possession Drift. These beds have the same structural attitude as the underlying Whidbey Formation and form part of one continuous northwest-dipping stratigraphic section that is more than 100 m thick and extends from the base of the coastal bluff to elevations of 15–20 m on the bluff face. Farther north between faults 2 and 3, sand and gravel outcrops (*m2* in fig. 32; not shown in fig. 27) have similar structural attitude, appear continuous with the underlying section, and are also correlated with the Possession Drift. The thickness of strata between faults 2 and 3 is not accurately portrayed in figure 32, which plots the distribution of strata relative to the elevation at which they occur.

The outcrops on opposite sides of the projection of fault 2 suggest a minimum of about 10 m of north-side-up vertical separation along the structure. Determining a more precise value for this separation is problematic for three reasons: (1) Uncertainty exists in the correlation of the section that occurs between faults 1 and 2; (2) Well data are lacking in the critical area just south of the fault (between wells 61 and 62 in fig. 32), thus the thickness of Possession Drift and the elevation of the top of the Whidbey Formation immediately south of fault 2 are unknown; (3) Projecting outcrop dips into the subsurface cannot be done with certainty. Cumulative vertical separation on faults 1 and 2 on the top of the Whidbey Formation thus appears to be a minimum of 20–25 m.

The projected trace of fault 3 is also covered. Three outcrops of Pleistocene strata occur between the inferred projections of faults 3 and 4 on Strawberry Point (figs. 27, 32). At the southern locality (*n1* in figs. 27, 32), a ≈4-m-thick section of interbedded sand and mud dips 24° NW. These strata are correlated with the Whidbey Formation based on lithology and sedimentology, and because they are at the same elevation as a

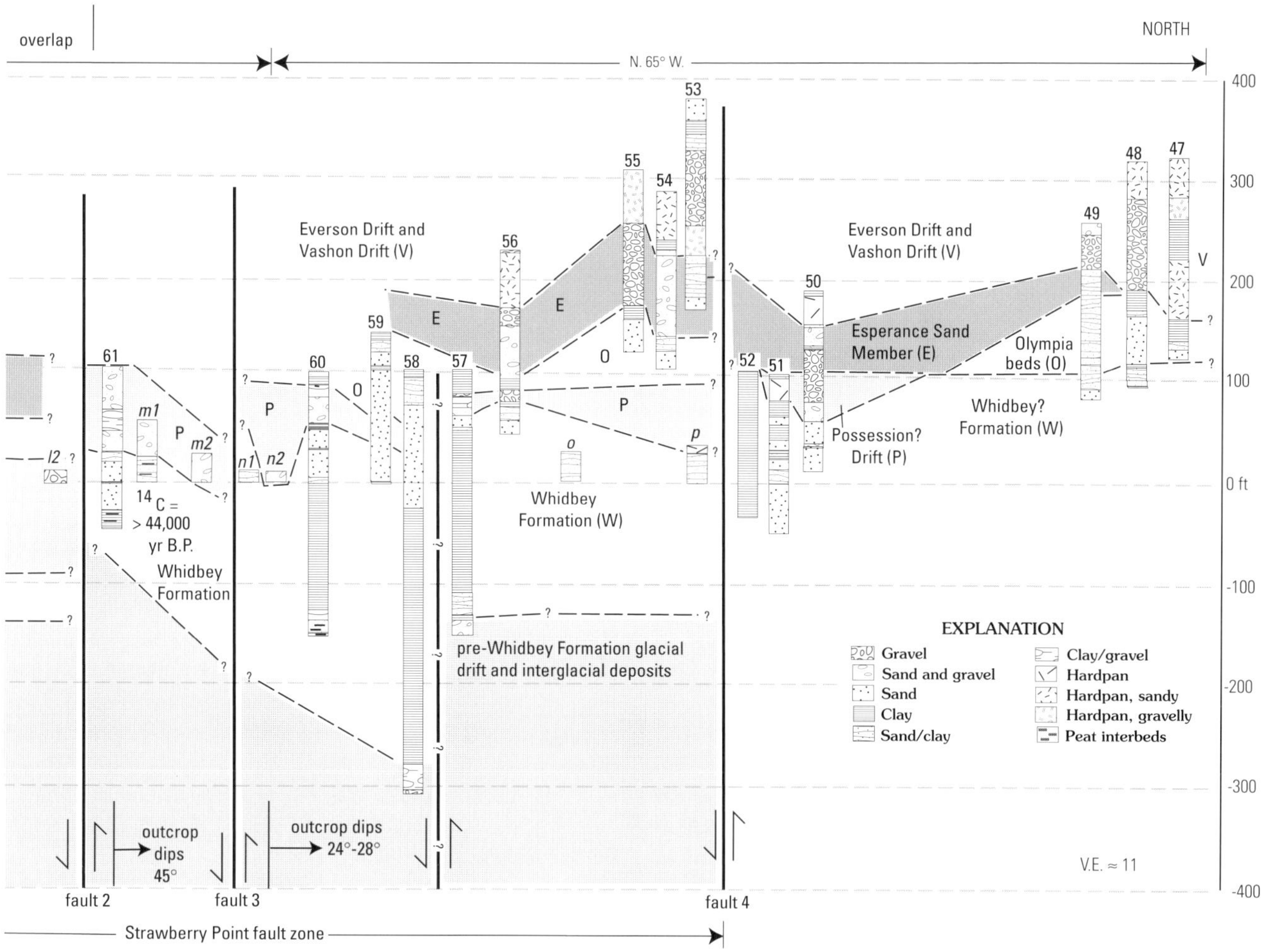

very thick (as much as 100 m) subsurface section of mainly clay and silt penetrated by nearby water wells (figs. 27, 28). The nonglacial Olympia beds are not known to reach comparable thickness anywhere in the region (for example, Dragovich and Grisamer, 1998; D.J. Easterbrook, written commun., 2000). A small sample of detrital wood from these strata yielded a radiocarbon date of 44,740±910 ^{14}C yr B.P., an age very close to the practical limit of radiocarbon dating. The sample could have been contaminated by minute amounts of younger carbon, and the date is here considered to be "infinite." Forty meters to the north (*n2* in fig. 32; not shown in fig. 27) across a covered interval, a 5-m-thick section of sand and gravel assigned to the Possession Drift, based on lithology and stratigraphic position, has a similar structural attitude. Thus, the Whidbey-Possession stratigraphic contact, exposed in the structural domain between faults 2 and 3, appears to be repeated between faults 3 and 4.

Determining the amount of vertical separation across fault 3 is problematic because of extensive cover between coastal exposures and the presence of just one useful well in the domain between faults 2 and 3. If one assumes that the contact between the Whidbey and Possession is repeated and that the structural dips in coastal exposures are continuous through the noncovered areas, then the amount of vertical separation needed to produce the repeated contact is about 80–120 m.

Outcrop data suggest that the section might be repeated not only across fault 3 but also between faults 3 and 4 by a fault not apparent on the seismic-reflection data. About 10 m of fractured, west-dipping (28°) sand and mud tentatively interpreted as Whidbey Formation is exposed at locality *o* (figs. 27, 31*D*); alternatively, these strata could have been deposited during the Olympia nonglacial interval and thus form the upper part of a continuous stratigraphic section, extending from the Whidbey Formation to Olympia beds, in the southern part of the domain between faults 3 and 4. The beds at *o* are cut by west-northwest-striking (≈70°) subvertical fractures.

Outcrop locality *p* occurs along the projected trace of fault 4 (fig. 27) and is the last bluff exposure on the northeast coast of Strawberry Point. The outcrop consists of 4–5 m of highly fractured friable sand (fig. 31*E*, *F*) overlain by pebbly diamict. Fractures have variable northwest to northeast strikes (50° to 340°), dip 30°–90°, are lined with clay, and commonly displace stratigraphic markers 2–10 cm. These strata are tentatively interpreted as uppermost Whidbey Formation overlain by Possession Drift. Well data suggest possible minor north-side-up slip on fault 4 (≈10 m) following deposition of the Whidbey Formation, but the amount is poorly constrained.

We interpret the faulting and folding within the outcrops at Strawberry Point to be tectonic in origin. This structural zone coincides precisely with the fault zone recognized a few hundred meters offshore on seismic-reflection data (figs. 5*A*, 8, 27). Similarly deformed Pleistocene strata are extremely rare in the Puget Lowland (among others, Gower and others, 1985; S.Y. Johnson, unpub. data, 1994–2000). Deformation by landsliding can be ruled out for several reasons: (1) Dipping sections are continuous vertically for as much as 100 m and laterally to the limits of exposure—they have not been disrupted or broken apart by gravity sliding, as is the case for many small recent landslides in the Strawberry Point area. (2) Dipping sections cross major stratigraphic contacts between mud- and sand-dominated units. These unit boundaries typically form the failure planes for Puget Lowland landslides. (3) The strike of dipping sections is perpendicular to the trend of the modern coastal bluffs, not parallel to the bluff free face as would be expected for latest Pleistocene or Holocene landslides. (4) Topography along Strawberry Point indicates the presence of many small landslides derived from the upper part of the bluff, but nothing remotely similar to the size of the observed dipping sections. Also, the dipping section between faults 2 and 3 extends to the top of the bluff; hence no source for a possible massive landslide exists in that location.

The coincidence with the offshore fault zone and the rare occurrence of deformed Pleistocene strata elsewhere in the Puget Lowland argue against a glaciotectonic origin (for example, Aber and others, 1989; Hart and Boulton, 1991; Hart and Roberts, 1994) for deformation in Strawberry Point outcrops. Beds are not deformed in the thin-skinned fold-and-thrust style that characterizes glaciotectonic deformation, and no obvious glaciotectonic landforms (such as composite ridges or push moraines) occur at Strawberry Point. Moreover, typical glaciotectonic features such as listric faults, small-scale folds, chevron folds, tectonic pods and boudins, augens, and pressure shadows are absent; secondary faults and fractures are generally subvertical and not low angle; and tectonic fabric is not widely distributed throughout the zone.

Utsalady Point Fault

West Coast of Whidbey Island

The Utsalady Point fault projects onto Whidbey Island's west coast across a sandy beach and lowland south of Rocky Point (figs. 27, 28), an area of no exposures. The closest exposures north of the fault are at Rocky Point (*g* in fig. 27), where pre-Tertiary basement rock is overlain by Everson Drift (fig. 6; Pessl and others, 1989). Easterbrook (1968) described the closest coastal bluff exposures to the south, 4 km away along bluffs south of Swantown (*d1* in fig. 27). These exposures consist of the Whidbey Formation overlain by Esperance Sand Member with no intervening Possession Drift or nonglacial Olympia beds (fig. 6).

Nearby offshore seismic-reflection profiles (figs. 10, 13) indicate about 200–300 m of down-to-the-south vertical separation on the base of the uppermost Pliocene(?) and Pleistocene section across the Utsalady Point fault. On land, this contact is at an elevation of a few meters above to 50 m below sea level north of the fault (fig. 33). No wells south of the fault (as deep as 300 m) encountered basement. The thicker Pleistocene section south of the fault appears to consist almost entirely of pre-Whidbey Formation glacial and interglacial deposits (fig. 33). The geometry of the Utsalady Point fault on offshore seismic-reflection data strongly suggests that the noted onshore relief on the base-of-uppermost Pliocene(?) to Pleistocene surface is created by a near-vertical fault bounding two subhorizontal surfaces and not by a gentle subsurface slope between Rocky Point and well 100 (figs. 28, 33). The top of basement also deepens to about 160 m about 1 km north of the Utsalady Point fault, indicating a possible south-side-up fault splay in this area, consistent with offshore seismic-reflection data (fig. 10).

Using the top of the interglacial Whidbey Formation as a younger datum (discussed in Devils Mountain fault subsurface data) for determining possible offset on the Utsalady Point fault is problematic on the figure 33 correlation diagram because of the apparent absence of Possession Drift on this part of Whidbey Island. In these wells, the top of the Whidbey Formation may be in contact with lithologically similar strata of the nonglacial Olympia beds or with the Esperance Sand Member as in the Swantown coastal bluff exposures (*d1*) at the south end of the profile. In either case, the top of the Whidbey Formation is not an obvious surface on lithologic logs of wells adjacent to the fault. Well data are thus consistent with Quaternary offset on the Utsalady Point fault in this area but do not provide the data needed to estimate the amount of post-Whidbey Formation offset.

Central and Eastern Whidbey Island

Across central and eastern Whidbey Island, the Utsalady Point fault is covered by vegetation and by surficial deposits developed mainly on Vashon Drift and recessional outwash (Pessl and others, 1989). No obvious topographic lineament is associated with the fault zone on existing topographic maps, and there is an insufficient number of deep water wells flanking the fault to construct stratigraphic correlation diagrams either parallel to or across the fault that might constrain location and slip. Thus, the location of the Utsalady Point fault as it crosses and possibly bifurcates on central and eastern Whidbey Island is speculative. Constraints on the location shown in figures 2, 27, and 28 include (1) the locations of faults on offshore seismic-reflection profiles (figs. 9, 10, 13), (2) weak aeromagnetic anomalies (fig. 3), (3) one deep test well south of the fault on the Maylor Peninsula (No. 101, fig. 28, table 1), which encountered bedrock at an elevation of about 475 m below sea level, and (4) a 500-m-long continuous exposure of flat-lying, locally fractured but unfaulted glacial and interglacial strata on the southeast coast of Polnell Point (*q* in fig. 27).

Camano Island

One splay in the central part of the Utsalady Point fault zone is exposed in the coastal bluffs at Utsalady Point on Camano Island (*r* in fig. 27). Strata are exposed almost continuously for 250 m along the coast and consist of three principal lithologies (figs. 34, 35): (1) laminated silty mud and sand with rare pebbles and organic detritus; (2) fine to medium sand and less common silty mud in which primary bedding has been largely destroyed by convolute stratification, pillar structures, and other soft sediment deformation features; and (3) lenses and wedges of gravel, pebbles, cobbles, and boulders. The geometries of units and the mix of lithologies suggest variable depositional environments within channels, floodplains, and possibly lakes of an alluvial plain, and postdepositional mixing of sandy strata by processes associated with liquefaction. Sparse fine-grained organic matter disseminated within sandy silt beds a few meters apart yielded radiocarbon dates of 21,100±150 and 15,190±220 ^{14}C yr B.P. (SJ-99-1A, 1B; table 2). The difference between the two dates is larger than expected given their close stratigraphic position. Both samples are low in carbon, and it is possible that the age of one or both samples is too young due to minor contamination.

These dates place the strata within the latter part of the Olympia nonglacial interval (fig. 6), and they indicate that these strata are younger than the Olympia beds exposed across Saratoga Passage on the southeast coast of Strawberry Point (fig. 27, table 2; Hansen and Easterbrook, 1974). In the Tacoma area of southern Puget Sound, Troost (1999) reported radiocarbon dates as young as 15,110 ^{14}C yr B.P. for Olympia beds. The older of the two ages from Utsalady Point is similar to many reported for the Coquitlam Drift, a pre-Vashon glacial formation exposed about 100 km to the north near Vancouver, B.C. (Hicock and Armstrong, 1981), and to the pre-Vashon Evans Creek Drift, deposited by valley glaciers extending from the Cascade Range to the edge of the Puget Lowland (Armstrong and others, 1965).

Beds at the Utsalady Point locality are deformed into a broad northwest-trending fold. Structural dip ranges from 24° (figs. 34, 35) on the northeast limb of the fold to 10° on the southwest limb. The fold axis is cut by a vertical east-southeast (120°) trending fault (figs. 33, 34*B*, *C*). This fault truncates irregular pods and wedges of massive sand and pebbles in the middle of an ≈8-m-wide zone in which primary bedding has been destroyed by faulting and soft-sediment deformation. The fault is characterized by a 20–30 cm wide zone of highly sheared sediment, which contains both rotated and broken clasts. Eight fractures adjacent to the master fault have a mean strike and dip of 127° and 84° SW., respectively.

Vertical offset on this fault appears to be less than a few meters based on the projected elevation of the contact between silty mud and sand facies on opposite sides of the mixed zone adjacent to the fault. The pebbly pods on the east side of the fault have probably been intruded from a source bed lower in the section. The source for the pebble wedge on the west side of the fault is not obvious; it may also have been intruded from lower in the section (requiring transport from out of the plane of the exposure), or alternatively, it could be the depositional fill of a large fissure that formed along the fault. Water-well data (fig. 37) show no nearby tills at this elevation, so formation and filling of this pebble wedge by nontectonic subglacial processes is unlikely.

In the bluff outcrops about 50 to 150 m west of this fault on the west limb of the fold, a 6-m-thick sandy horizon displays large-scale soft-sediment deformation (fig. 35*D*). Features include numerous dikes and pillar structures, convolute stratification, ball-and-pillow structure, and intrusion by large dikes of sandy silt to gravel that extend up through as much as 13 m of section. Subvertical fractures with a mean strike of 305° (n=16) are common on the west limb of the fold.

As with the Pleistocene strata exposed at Strawberry Point (see p. 23–30, Strawberry Point fault, east coast of Whidbey Island), we infer that the faulting and folding of beds at Utsalady Point is tectonic in origin. The coincidence with the fault zone recognized offshore in Saratoga Passage is striking (fig. 9), and such deformation of Pleistocene strata in the Puget Lowland is extremely rare (Gower and others, 1985; S.Y. Johnson, unpub. data, 1994–2000). Landsliding or glaciotectonic origins can be ruled out for most of the same reasons cited for the Strawberry Point fault.

Two faults are exposed in a vegetated roadcut 900 m southeast (128°) of the fault exposed at Utsalady Point (*s* in fig. 27). Strata cut by these faults consist of sand and gravel and are inferred to be Vashon glaciofluvial advance deposits, correlative with the Esperance. One fault (fig. 36*A*) is subvertical, strikes 95°, juxtaposes massive gravel (on the south) and plane-stratified sand (on the north), and has minimum vertical displacement of 150 cm. The second fault (fig. 36*B*) occurs 15 m to the southeast, strikes 0° and dips 67° E., juxtaposes massive gravel (on the east) and stratified sand (on the west), and similarly has minimum vertical displacement of 150 cm. These exposed faults indicate southeastward continuation of the Utsalady Point fault zone, consistent with the trend of aeromagnetic anomalies (fig. 3).

Figure 37 shows a stratigraphic correlation diagram across part of the Utsalady Point fault zone based on lithologic logs from water wells, with the bluff exposure and dates for the Olympia beds at Utsalady Point (figs. 27, 34, 35; location *r*) providing important stratigraphic control. The outcrop section at *r* correlates with a fine-grained interval in the subsurface that is much thicker than any known section of Olympia beds from this part of the Puget Lowland. This suggests that the Possession Drift is missing from this local area and that Olympia beds directly overlie the lithologically similar interglacial deposits of the Whidbey Formation. Because of this lithologic similarity, there is no basis for inferring the Whidbey-Olympia contact in the subsurface near this locality, and no obvious stratigraphic marker is present to constrain vertical offset on the fault exposed at Utsalady Point.

The southwestern splay in the Utsalady Point fault zone projects onto the correlation diagram between wells 106 and 107 (figs. 28, 37). The stratigraphy of wells southwest of this projected fault differs from those to the northeast in several ways. (1) Pre-Whidbey Formation glacial drift is inferred to be present at the bottom of one deep well on the southwest side of the fault but was not encountered at similar depths by wells northeast of the fault. (2) The Possession Drift is inferred to be present in five of six wells southwest of the fault; the Possession is apparently absent in the five wells northeast of the fault. (3) West of the fault, the elevation of the inferred top of the Whidbey Formation is more than 15 m; east of the fault it is below sea level at outcrop locality *r* (figs. 27, 34, 35, 37). (4) The top of the Olympia beds in six wells southwest of the fault has a mean and maximum elevation of 56.7±7.9 m and

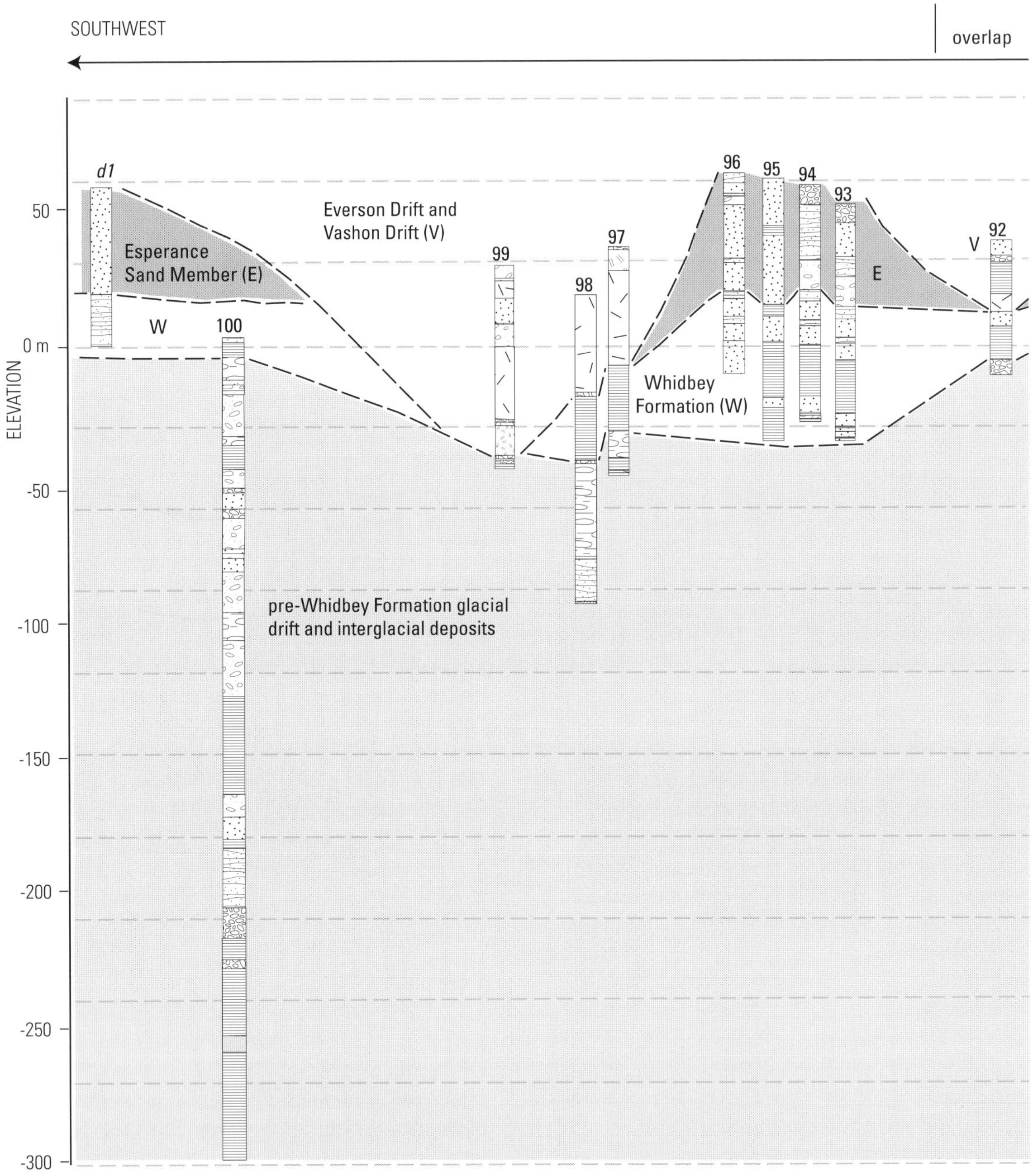

Figure 33. Interpretive stratigraphic correlation diagram crossing Utsalady Point fault zone, northwestern Whidbey Island (figs. 27, 28). Correlation diagram is based on a few outcrops (fig. 27; lettered sections) and lithologic logs from water wells (table 1, fig. 28; numbered sections). Faults dashed where inferred, queried where extent uncertain; barbs show direction of relative fault movement. Note overlap between the two panels of this figure.

66.4 m in contrast to 26.9±3.8 m and 30.8 m for three wells northeast of the fault. This difference is much less, only 11.3 m, in the two wells closest to and on opposite sides of this fault (106 and 107). These contrasts cumulatively suggest that the southwest side of the fault has moved up relative to the northeast side. This is opposite to the sense of throw along the Utsalady Point fault on western and central Whidbey Island. Vertical offset on this fault is not obvious on the nearby (≈2.5 km) offshore seismic-reflection profile in Saratoga Passage (figs. 5, 9).

Fault Mapping, Style, and Rates of Deformation

Devils Mountain Fault

Offshore seismic-reflection and onshore geologic data indicate that the Devils Mountain fault is continuous from the Cascade Range foothills (Whetten and others, 1988; Tabor and others, 1988, Tabor, 1994) across northern Whidbey Island and into the eastern Strait of Juan de Fuca, a minimum distance of approximately 125 km (fig. 1). The Devils Mountain fault was

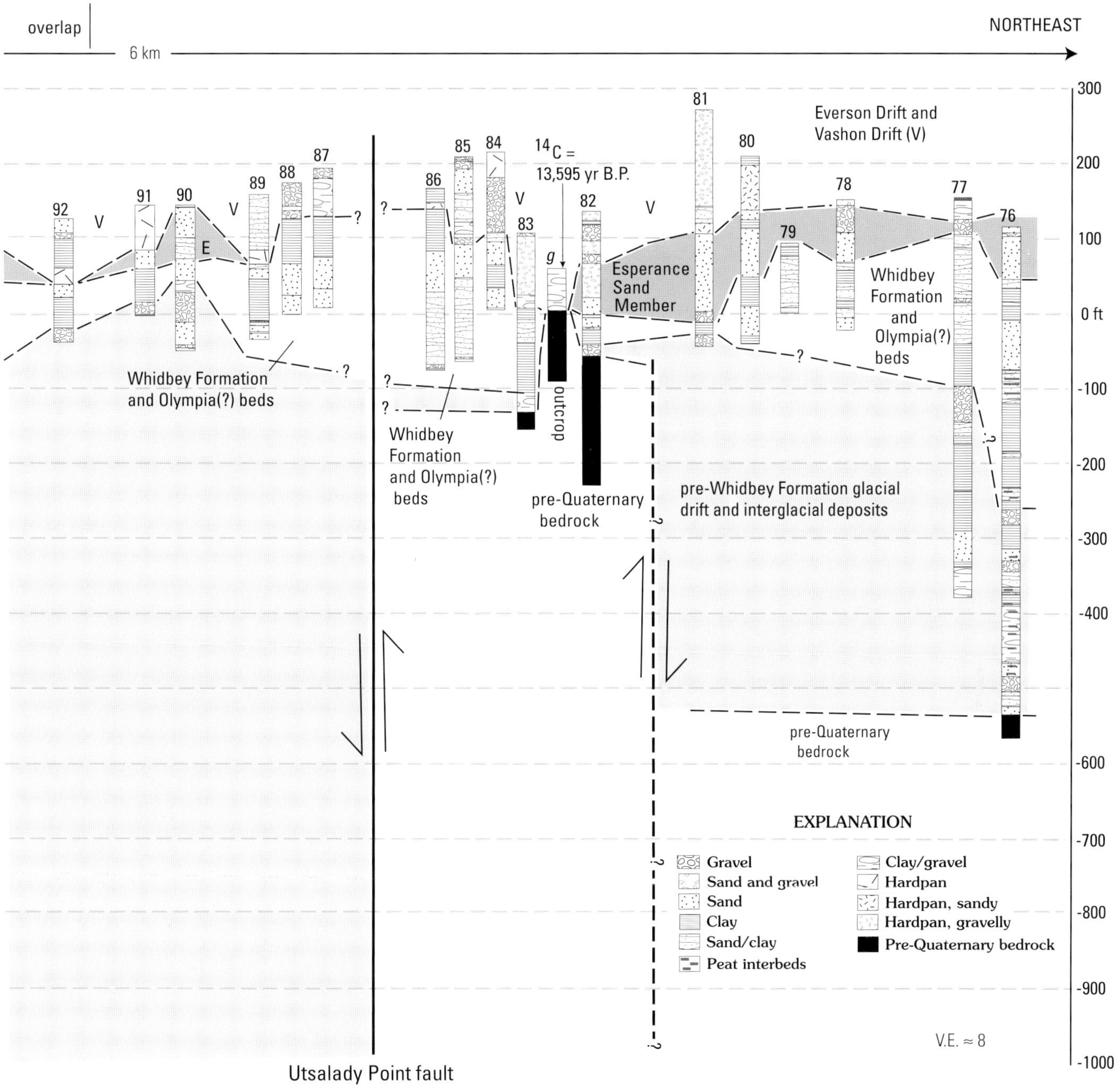

recognized on all but one of 14 crossing high-resolution seismic-reflection profiles and all deeper industry profiles, and it appears to cut and (or) fold inferred Quaternary reflections on virtually every profile (figs. 8–26). The dip of the upper 1–2 km of the fault plane varies from about 45° to 75° (61°±10° for crossing seismic profiles) to the north with the steepest dips to the east in profiles flanking Whidbey Island. The pattern of juxtaposed reflections on some profiles also suggests out-of-plane (strike-slip) offsets.

For its entire length, the fault coincides closely with the southern margin of a variable-amplitude aeromagnetic anomaly (fig. 3). The variability in amplitude is inferred to correlate mainly with the lithology of basement rocks, being strongest where pre-Tertiary igneous basement occurs and weakest where the basement consists of pre-Tertiary and Tertiary sedimentary rocks (fig. 6). The fault also roughly corresponds with a change in the depth to the base of uppermost Pliocene(?) to Pleistocene strata (fig. 7).

Seismic-reflection and aeromagnetic data east of Whidbey Island in Skagit Bay (figs. 2, 3, 8) suggest at least one tear fault or transfer zone along the Devils Mountain fault that could represent a segment boundary. A second segment boundary could be present in the eastern Strait of Juan de Fuca south of southeastern San Juan Island. In this area, the westernmost USGS high-resolution seismic-reflection profile (fig. 5*A*, west of the area of fig. 20) did not image a fault zone on strike with the Devils Mountain zone to the east and west. Although a nearby industry profile (fig. 21) did image a fault along the strike of the

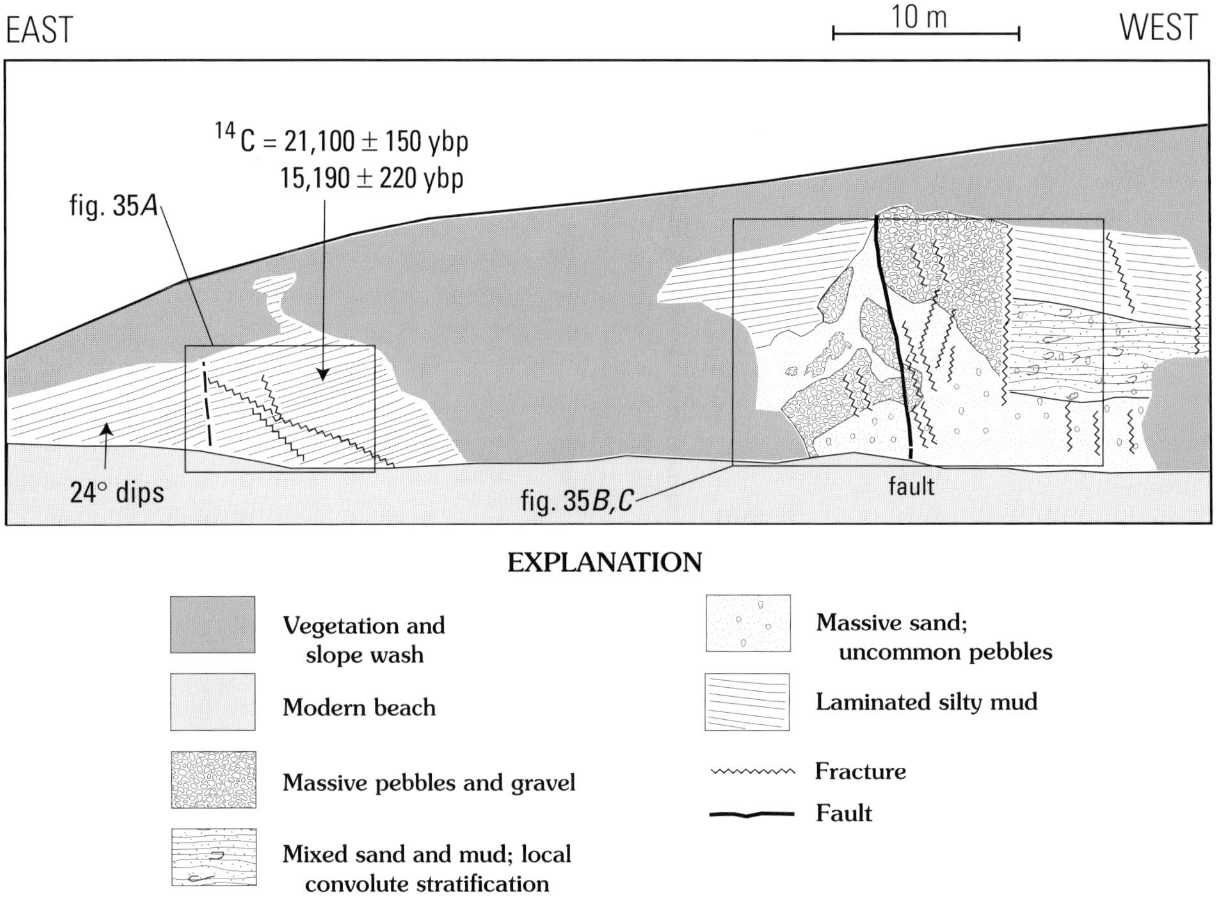

Figure 34. Line drawing (traced from photo mosaic) of deformed Quaternary strata exposed at location *r* (fig. 27), Utsalady Point, Camano Island.

Devils Mountain fault, this structure does not appear to deform the Quaternary section and may be an inactive strand. Other possible segment boundaries or gaps may be present along the fault trace but are presently unrecognized.

North of the Devils Mountain fault in the eastern Strait of Juan de Fuca, seismic-reflection data indicate three zones of northwest-trending folds and faults (fig. 2). Mapping these structures is limited by the incomplete coverage by seismic-reflection profiles (fig. 5); however, their trends correspond closely with aeromagnetic anomalies (fig. 3). Figure 2 shows their minimum lengths and extent. The eastern zone occurs at the south end of Rosario Strait and coincides closely with a high-amplitude aeromagnetic anomaly (fig. 3) that forms the southern boundary of a Jurassic ophiolite (Whetten and others, 1980). These faults and folds are well imaged on seismic-reflection data collected by Puget Power (1979). The middle zone of northwest-trending faults and folds occurs south of western Lopez Island and correlates with low-amplitude aeromagnetic anomalies that extend northwest, parallel to the southwest coast of San Juan Island (figs. 2, 3). The gradients of these magnetic anomalies indicate sources deeper than the section imaged by the seismic data, but it appears that the structure of these deeper sources is controlling the structural pattern of the upper 1 km. The western zone occurs at the mouth of Haro Strait and coincides closely with moderate- to high-amplitude aeromagnetic anomalies (figs. 2, 3). Many of the northwest-trending structures in these three zones cut and (or) fold reflections in inferred Quaternary strata. Thus, these faults and folds appear to be Quaternary en-echelon structures active at the same time as the Devils Mountain fault and not older, inactive features cut by a younger Devils Mountain fault.

The Devils Mountain fault probably continues west of our data base, passing a few kilometers south of the city of Victoria and merging to the west with the Leech River and (or) the San Juan fault on southern Vancouver Island (Muller, 1983). At present, however, neither the Leech River nor San Juan fault has been recognized as an active or potentially active fault. An alternative is that much or all of the deformation along the fault is transferred to the presumed en-echelon northwest-trending structural zone that coincides with the prominent aeromagnetic anomaly in Haro Strait (figs. 2, 3) which extends onto Vancouver Island north of Victoria. Geological Survey of Canada seismic-reflection Line 35 (fig. 26) is important in distinguishing between these two end-member alternatives. This line crosses the Devils Mountain fault south of the northwest-trending structural zone in Haro Strait and shows significant deformation of the base of the uppermost Pliocene(?) to Pleistocene section on the west-trending fault trace.

The map-view geometry of the Devils Mountain fault and associated structures resembles that described for many left-lateral, oblique-slip fault zones (for example, Christie-Blick and Biddle, 1985), with the Devils Mountain fault acting as the "master fault" within a broadly distributed zone of deformation.

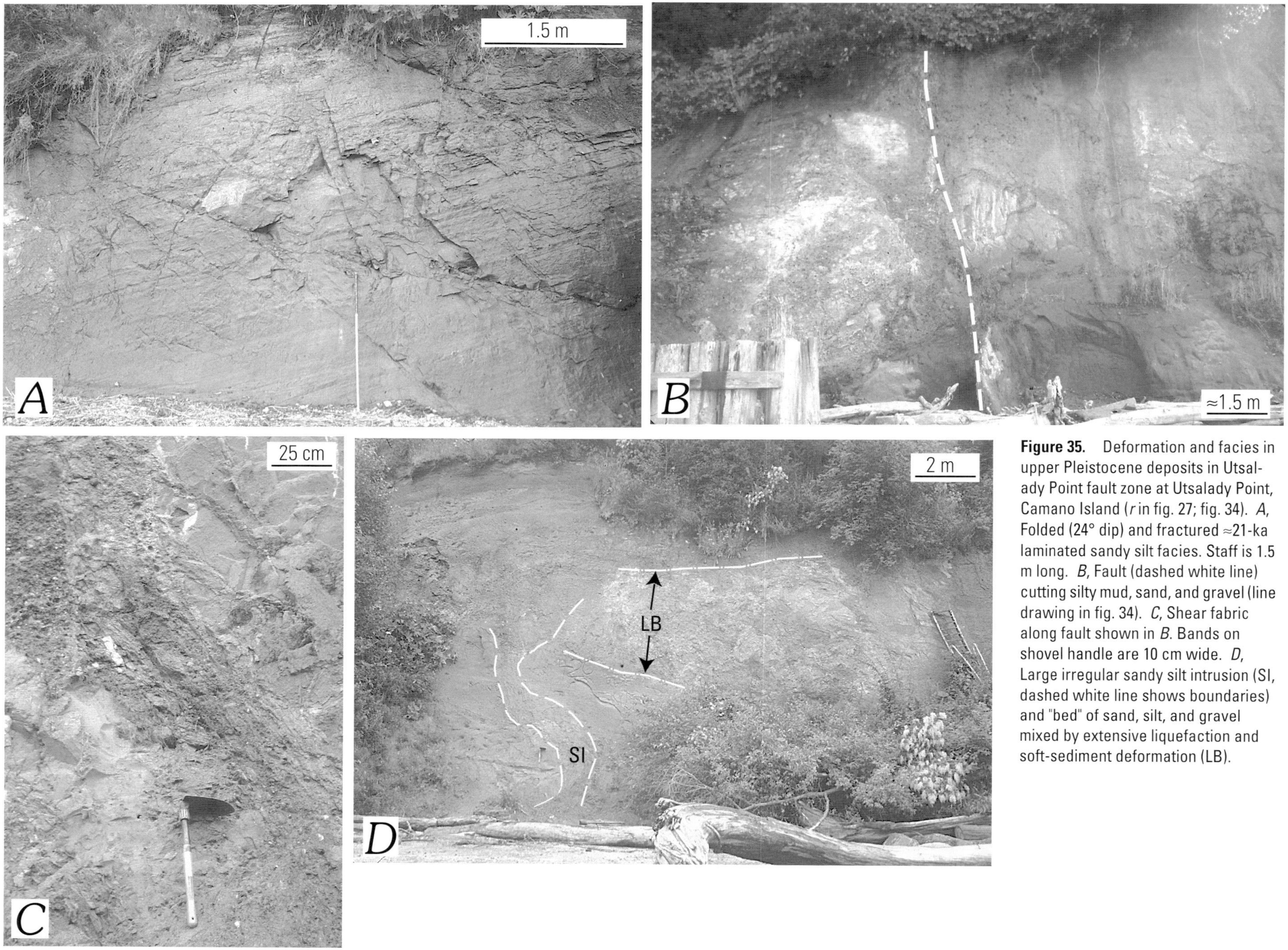

Figure 35. Deformation and facies in upper Pleistocene deposits in Utsalady Point fault zone at Utsalady Point, Camano Island (*r* in fig. 27; fig. 34). *A*, Folded (24° dip) and fractured ≈21-ka laminated sandy silt facies. Staff is 1.5 m long. *B*, Fault (dashed white line) cutting silty mud, sand, and gravel (line drawing in fig. 34). *C*, Shear fabric along fault shown in *B*. Bands on shovel handle are 10 cm wide. *D*, Large irregular sandy silt intrusion (SI, dashed white line shows boundaries) and "bed" of sand, silt, and gravel mixed by extensive liquefaction and soft-sediment deformation (LB).

Figure 36. Faults (*A, B*) exposed in a vegetated roadcut 900 m southeast of fault exposed at Utsalady Point (*s* in fig. 27), cutting inferred Vashon glaciofluvial advance deposits. The faults, which juxtapose well-stratified sand and massive gravel, are 15 m apart and occur within the Utsalady Point fault zone.

This map geometry notably contrasts with the expected structural pattern across a zone of normal contraction, a setting in which secondary faults and folds would have trends similar to that of a master fault. Based on the distinctive map pattern, we believe that the Devils Mountain fault is a transpressional, oblique-slip fault, characterized by both north-south shortening and left-lateral slip.

The vertical component of slip on the Devils Mountain fault can be constrained with onshore subsurface data (fig. 30) from Whidbey Island, assuming that stratigraphic interpretation of the top of the Whidbey Formation is correct and that similar amounts of postdepositional modification of this surface have occurred on each side of the fault. Given the interpretation in figure 30, the vertical separation on the top of the Whidbey Formation (80 ka; fig. 6) across the fault is about 11.9±8.3 m. This offset requires between about 4 and 24 m of fault offset (12–14 m for the mean), assuming (based on nearby seismic-reflection profiles) the fault dips 60°–75° in this area. The result is a late Pleistocene vertical slip rate of about 0.05–0.30 mm/yr (0.15–0.18 mm/yr for the mean) and a shortening rate of 0.01–0.15 mm/yr (fig. 38).

Glaciomarine features provide another possible tool for constraining fault offset. Dethier and others (1995) noted a ≈10 m inflection point (up to the north) coinciding with the Devils Mountain fault on a graph showing elevation of a ≈13 ka glaciomarine surface representing the maximum altitude of the marine limit. These data are consistent with the north-side-up offset on the Devils Mountain fault determined in this investigation. At present, however, data points for such glaciomarine features are too widely spaced to provide valid constraints for the Devils Mountain fault and other local structures. New high-resolution topographic imaging tools such as airborne laser terrane mapping (for example, Kitsap County Public Utility District, 1998; Bucknam and others, 1999; Nelson and others, 1999; Harding and Berghoff, 2000) may provide the best opportunity for the type of detailed investigations needed to use glacial and glaciomarine landforms in studies of active faults in this region.

The vertical component of slip on the Devils Mountain fault can also be estimated using offshore high-resolution seismic-reflection data. The drawback to this approach is that multiple cycles of widespread Quaternary glacial erosion and deposition in the eastern Strait of Juan de Fuca have affected both bedrock and Pleistocene deposits. As a result, the age of the lowest preserved uppermost Pliocene(?) to Pleistocene deposits on any one profile is unknown, and correlation of Quaternary units across the fault zone is either difficult or impossible. The most conservative approach is to use the base of the uppermost Pliocene(?) to Pleistocene section as a datum and to assign this surface a maximum age of 2 Ma (see section, "Identification and Age of the Base of the Uppermost Pliocene(?) to Pleistocene Section"). Using this datum on six profiles (figs. 8, 10, 13, 16, 20, 26) yielded minimum Quaternary slip rates varying from 0.03 to 0.13 mm/yr (fig. 38). We applied the methods of Schneider and others (1996, their fig. 10) to Geological Survey of Canada Line 37 (fig. 25), a special case in which a fault-propagation fold is imaged above the locally blind Devils Mountain fault. The vertical and horizontal components of fold growth are 180 m and 135 m, respectively, and inferred uppermost Pliocene(?) to Pleistocene fault

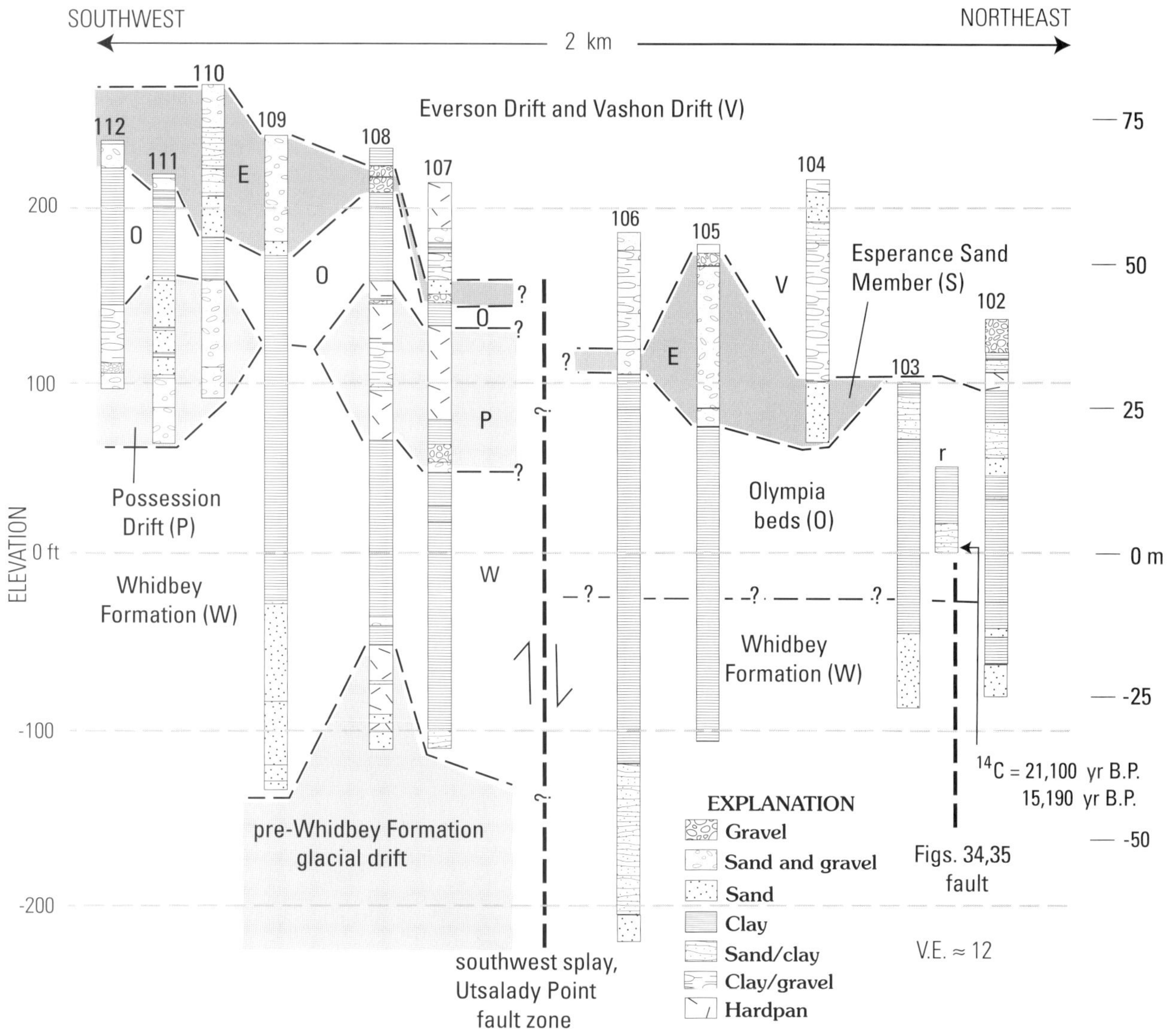

Figure 37. Interpretive stratigraphic correlation diagram crossing part of Utsalady Point fault zone at Utsalady Point, northwestern Camano Island. Correlation diagram is based on lithologic logs from water wells (table 1, fig. 28; numbered sections) and one outcrop (*r* in fig. 27; figs. 34, 35). Faults dashed where inferred, queried where extent uncertain; barbs show sense of relative offset.

displacement is 225 m on a plane dipping 53°. Using 2 Ma as the base of the Quaternary section yields an approximate minimum vertical slip rate of about 0.11 mm/yr.

The longer term minimum slip rates determined from seismic-reflection data are thus consistent with the shorter term, approximate, late Pleistocene vertical-slip rate (0.05–0.30 mm/year) estimated from onshore subsurface data (fig. 38). We emphasize that this rate reflects only the vertical component of slip on a structure we infer to be an oblique-slip fault.

Strawberry Point Fault

Seismic-reflection data and local onshore geology suggest that the west-trending Strawberry Point fault is continuous from the eastern Strait of Juan de Fuca to at least Skagit Bay, a minimum distance of 22 km (figs. 2, 27). It occurs as one discrete subvertical structure with its south side up along western Whidbey Island and in the easternmost Strait of Juan de Fuca (figs. 10, 13), where it forms the northern boundary of a relatively nonmagnetic basement block (fig. 3). The pre-Tertiary sedimentary rocks that form the core of this block are exposed at Rocky Point on the west coast of Whidbey Island (fig. 2). We infer about 80–200 m of vertical relief on the base of the uppermost Pliocene(?) to Pleistocene section along the Strawberry Point fault on two seismic-reflection profiles (figs. 10, 13) west of Whidbey Island. Assuming a maximum age of 2 Ma for this contact (see section, "Identification and Age of the Base of the Uppermost Pliocene(?) to Pleistocene Section"), this relationship yields a minimum vertical slip rate for the Quaternary of 0.04–0.10 mm/yr (fig. 36). The occurrence of late Holocene (≈1.7 ka) bulrush rhizomes at sea level on the south side of the fault at Rocky Point (fig. 27; table 2) provides a horizontal datum that rules out south-side-up vertical offset of more than a few meters on the Strawberry Point fault in the last 1,750 years.

Onshore and offshore Strawberry Point on eastern Whidbey Island, deformation is distributed on several subvertical splays and gentle to moderate folds in a ≈2-km-wide zone. Inferred

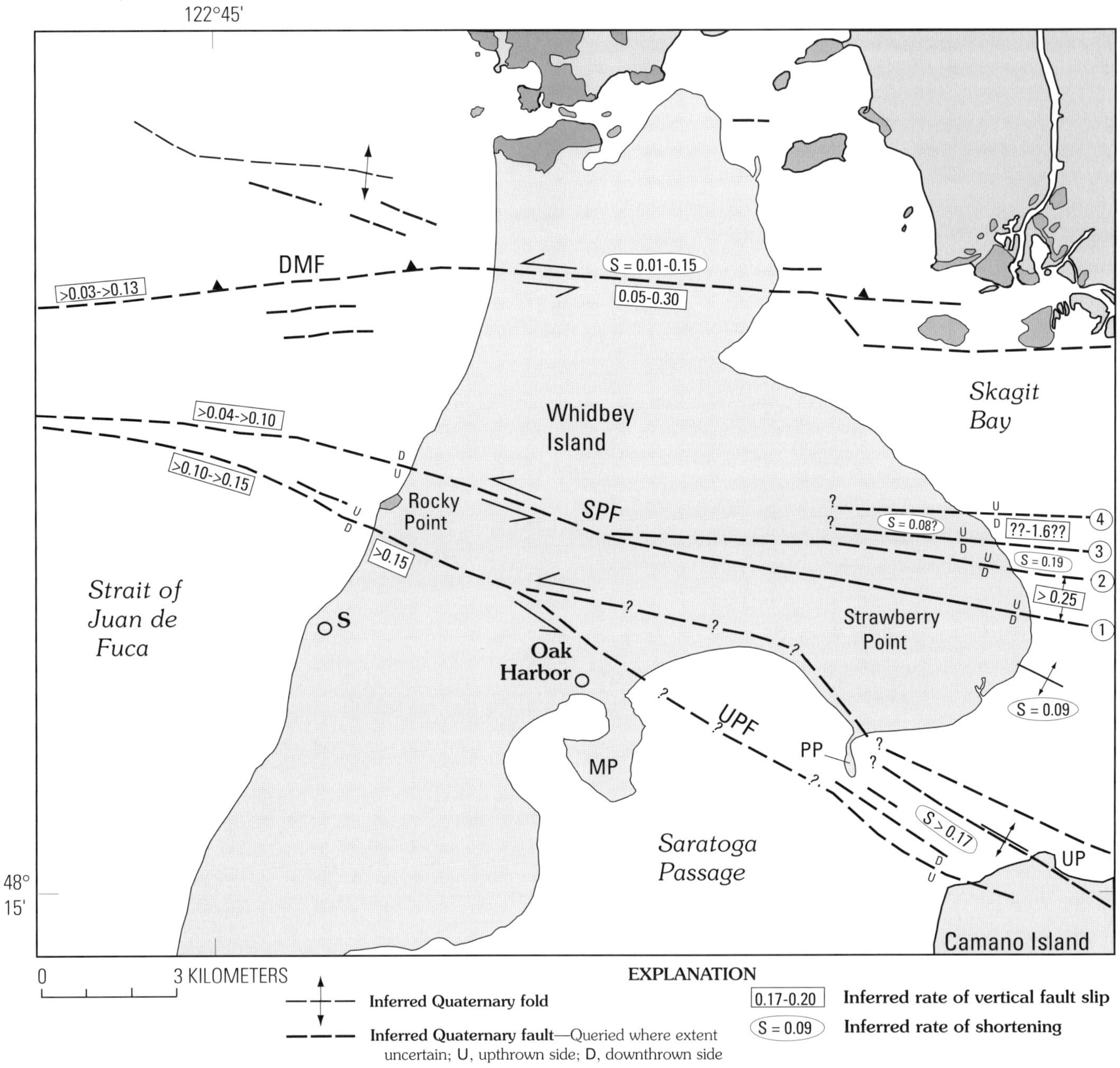

Figure 38. Northern Whidbey Island area showing locations of faults and folds that deform Quaternary deposits, inferred rate (commonly a minimum) of vertical fault-slip (mm/yr), and shortening rates on the Devils Mountain fault (DMF), Strawberry Point fault (SPF), and Utsalady Point fault (UPF). Queried rates are based on meager data and are highly speculative, and each fault probably has a significant component of lateral slip for which slip-rate information is not available. Dark-shaded areas are underlain by pre-Quaternary bedrock (fig. 6). Light-shaded areas are underlain by Quaternary deposits. MP, Maylor Point; PP, Polnell Point; S, Swantown; UP, Utsalady Point.

stratigraphic offsets (fig. 32) suggest that the four main subvertical splays have north-side-up displacement, opposite to the sense of vertical fault slip on the Strawberry Point fault west of Whidbey Island. Using the top of the Whidbey Formation as a datum, the fold south of fault 1 (fig. 27) has a late Pleistocene uplift rate of about 0.38 mm/year (fig. 37). The shortening rate required to produce this uplift is about 0.09 mm/yr (method of Rockwell and others, 1988). The cumulative minimum latest Quaternary (80 ka to present) vertical slip rate for faults 1 and 2 is inferred to be about 0.25 mm/yr given the favored, more conservative stratigraphic correlation. The folded strata between faults 2 and 3 require an uplift rate of 0.88 mm/yr and a shortening rate of 0.19 mm/yr.

The amount of vertical slip on faults 3 and 4 is not well constrained. Outcrops of upper Pleistocene strata locally dip 24° to 48° (fig. 27), making projections of stratigraphic surfaces tenuous. Making the speculative assumption that dips and stratigraphic contacts are continuous yields vertical slip rates as high as 1.0–1.5 mm/yr on fault 3 and about 0.1 mm/yr on fault 4. Shortening rates recorded by folded strata between faults 3 and 4 are likewise difficult to estimate because of poor outcrops; our poorly constrained best estimate is ≈0.08 mm/yr.

The cumulative vertical slip rate on the four splays of the Strawberry Point fault at Strawberry Point is thus inferred to be a minimum of about 0.25 mm/yr, but larger rates are probable. The cumulative shortening rate recorded by folded strata within and adjacent to the fault zone is a minimum of about 0.35 mm/yr.

Folding in outcrops within and adjacent to the Strawberry Point fault zone at Strawberry Point indicate contractional deformation that, together with the reversal of the sense of vertical offset in the zone across Whidbey Island, strongly suggests the Strawberry Point fault is an oblique-slip transpressional fault (Christie-Blick and Biddle, 1985). Regional fault geometry (see "Fault Mapping, Style, and Rates of Deformation"—"Devils Mountain Fault") suggests left-lateral displacement, but we have not identified piercing points that could be used to quantify the sense, amount, and rate of inferred lateral slip.

Utsalady Point Fault

Seismic-reflection profiles and local onshore geology suggest that the northwest-trending Utsalady Point fault is continuous from the eastern Strait of Juan de Fuca across Whidbey Island, Saratoga Passage, and Camano Island, a minimum distance of 28 km (fig. 2). East of Whidbey Island, the projection of the Utsalady Point fault zone coincides with the southern margin of a low-amplitude aeromagnetic high (fig. 3; Blakely and others, 1999b). This anomaly extends at least 25 km southeast of Utsalady Point, suggesting that the fault may have a length of 50 km or more.

The Utsalady Point fault occurs as one discrete subvertical structure with north-side-up offset west of Whidbey Island (figs. 2, 10, 13, 27, 28). We infer about 200–300 m of vertical relief across the fault at the base of the Quaternary section on two seismic-reflection profiles (figs. 10, 13) in this area. Assuming a maximum age of 2 Ma for this contact (see section, "Identification and Age of the Base of the Uppermost Pliocene(?) to Pleistocene Section"), this displacement yields a minimum vertical slip rate for the Quaternary of about 0.10–0.15 mm/yr (fig. 37). Onshore western Whidbey Island, subsurface data (fig. 33) indicate more than 300 m of relief on the base of the Quaternary, yielding a minimum vertical slip rate of 0.15 mm/yr since about 2 Ma.

Seismic-reflection data from Saratoga Passage (fig. 9) east of Whidbey Island represent the fault as a broad (≈1.5 km) zone comprising several splays, on which significant vertical offset on the base of the Quaternary is not obvious. Camano Island onshore data (figs. 34–37) suggest that the southernmost splay in the zone has north-side-down offset, and therefore that the sense of vertical displacement changes along the strike of the fault. This along-strike variation, together with the evidence for contractional deformation in the fault zone (folding of ≈21 ka beds; figs. 34, 35), suggests that the Utsalady Point fault is also an oblique-slip transpressional structure (Christie-Blick and Biddle, 1985). As discussed (see "Fault Mapping, Style and Rates of Deformation"—"Devils Mountain Fault"), regional fault geometry suggests left-lateral displacement, but we know of no piercing points that can be used to quantify the sense, amount, and rate of inferred lateral slip.

Subsurface data from Camano Island (fig. 37) do not provide sufficient information on the top of the Whidbey Formation for it to be an effective marker in constraining vertical offset along the Utsalady Point fault. As noted previously ("Evidence for Faulting Onshore, Whidbey and Camano Islands"), the top of the Olympia beds is significantly higher on the south side of the southern splay in the zone, but its elevation appears to drop to the north approaching this splay, and this datum is only 11 m higher on the south than to the north in the two wells (106, 107) closest to the splay. Moreover, seismic-reflection data (fig. 9) do not show obvious south-side-up offset, and considerable erosional relief occurs on the top of the Olympia beds on Whidbey Island and elsewhere in the Puget Lowland (Troost, 1999, among others). Regardless of uncertainties concerning stratigraphic correlation and vertical slip rate in the Camano Island region, outcrops at Utsalady Point indicate significant post-20 ka deformation in the Utsalady Point fault zone. The folded strata at Utsalady Point record a minimum uplift rate of 0.81 mm/yr and a shortening rate of 0.17 mm/yr.

Discussion

The geometry of structures in the eastern Strait of Juan de Fuca–northern Puget Lowland region suggests that the Devils Mountain fault is a transpressional, left-lateral, oblique-slip "master fault," and that the Strawberry Point fault and Utsalady Point faults are probably en-echelon structures. This deformation may be driven by stress associated with convergence along the continental margin. GPS data from this region (Dragert and Hyndman, 1995; Khazaradze and others, 1999) indicate that this stress produces significant east-northeast-directed crustal strain subparallel to the plate convergence vector. The larger part of this strain is thought to be elastic, recoverable following a large plate-boundary earthquake. A smaller component might result in inelastic crustal faulting and folding. In this setting, left-lateral slip would be favored on east-west-trending structures such as the Devils Mountain fault.

Although there are no obvious piercing points in available data that can be used to quantify the amount and rate of lateral slip across this zone, regional geology provides possible constraints. Whetten (1978) suggested about 60 km of left-lateral slip on the Devils Mountain fault based on offset of the Trafton sequence, a distinctive Mesozoic melange unit (part of the eastern melange belt of Tabor, 1994), using piercing points in the San Juan Islands north of the fault and Cascade Range foothills south of the fault. This potential piercing point is suspect because many workers (such as Whetten and others, 1980; Brandon and others, 1988; Tabor, 1994) have shown that the Trafton unit occurs within a stack of thrust sheets that might have had greater lateral extent than present exposures demonstrate. Also, this unit could occur farther west than present exposures reveal on the south side of the fault in the subsurface of the Everett basin. This possible piercing point does, however, probably reveal the maximum amount of left-lateral slip possible across the zone.

The first obvious record of movement along the Devils Mountain fault dates from about 40 Ma, when coarse (clasts as large as 1 m) alluvial-fan conglomerate of the rocks of Bulson Creek was deposited adjacent to the fault in the Everett basin (Marcus, 1980; Whetten and others, 1988). These conglomerates include silicic volcanic clasts that are petrographically similar to Eocene ash-flow tuff units that crop out north of the fault, apparently limiting the amount of post-40 Ma left-lateral slip to less than about 30 km. Making the highly questionable assumptions that (1) the rate of left-lateral slip has been constant over the last 40 m.y. and (2) 30 km is the maximum amount of left slip possible across the structure yields a maximum lateral slip rate of 0.75 mm/yr over this time period. There are no similar questionable constraints for the possible amount of lateral slip across the Strawberry Point and Utsalady Point faults.

Vertical slip rates on the Strawberry Point and Utsalady Point faults appear comparable to or higher than those on the Devils Mountain fault (fig. 37), the structure that regional fault geometry suggests has been the "master fault" within the system. These rates and the evidence for post-80 ka deformation from outcrops at Strawberry Point and Utsalady Point (figs. 31, 35) may indicate that fault kinematics have changed or are in a transitional period in which local deformation has most recently been concentrated on the Strawberry Point and Utsalady Point faults rather than the Devils Mountain fault. Such changes are common within complex strike-slip fault systems (Christie-Blick and Biddle, 1985) and are not surprising. The map geometry of the Strawberry Point and Utsalady Point faults also indicates that they may be connected at depth and represent components of a large transpressional flower structure (Harding and others, 1983; Christie-Blick and Biddle, 1985).

Given the evidence for left-lateral slip on the Devils Mountain, Strawberry Point, and Utsalady Point faults, we believe that deformation along the northwest-trending southern Whidbey Island fault to the south in the eastern Strait of Juan de Fuca region (fig. 1) is mainly contractional. This interpretation is consistent with seismic-reflection data from the eastern Strait of Juan de Fuca presented in Johnson and others (1996, their fig. 5), who showed this structure as a zone of thrust faults. Johnson and others (1996) previously suggested a component of right-lateral slip on the southern Whidbey Island fault because it forms a part of the crustal boundary on the east and northeast flank of the north-moving Coast Range block. Right-lateral slip on this crustal boundary is required farther to the south.

Determining the rate of lateral slip across the structural zone comprising the Devils Mountain, Strawberry Point, and Utsalady Point faults is thus an important problem left unresolved by this investigation. The difficulty in constraining lateral slip across this structural zone in large part reflects the recent glacial and post-glacial (≈20 ka) age of the landscape. Landforms of tectonic origin (such as scarps, pressure ridges, terraces, or offset drainages) have been eroded or filled in by repeated pulses of glacial erosion and deposition so that, to a large degree, only landforms generated during or after the last glaciation are evident. Given the rates described previously and the relatively friable nature of many glacial deposits, post-glacial landforms may be subtle and have limited preservation potential.

New high-resolution topographic and bathymetric imaging tools may provide the best opportunity for identification of tectonic landforms in this landscape. For example, Bucknam and others (1999) and Nelson and others (1999) used a high-resolution airborne laser terrane mapping survey to identify a post-glacial scarp in a dense forest in the Seattle fault zone. New multibeam bathymetric surveys from offshore California (Bohannon and others, 1998, among others) have shown many tectonic features and offsets associated with strike-slip faulting in the seafloor, and such surveys could have similar value in the eastern Strait of Juan de Fuca.

Earthquake Hazards

Earthquake Occurrence

The historic earthquake record and recorded seismicity support the interpretation that this region is seismically active. The historic record extends back to about 1850 but is of variable quality until about 1929 (Ludwin and Qamar, 1995); accurate recording of regional earthquakes began in 1969 when the Pacific Northwest Seismograph Network was installed. It is noteworthy that Victoria (fig. 1) felt a cluster of at least 14 earthquakes between 1859 and 1870 with two having a Mercalli Magnitude Index of VI. The source of these earthquakes is unknown, but they might be associated with the nearby Devils Mountain fault.

Recorded crustal earthquakes in the study area with magnitude (M) larger than 2.5 are shown in figure 2. Between Haro Strait and Lake Cavanaugh (figs. 1, 2), these earthquakes and smaller events have clustered at depths of 12–20 km along a 50° N. dip projection from the surface trace of the Devils Mountain fault (Stanley and others, 1999). If these events are associated with the Devils Mountain fault, then its north dip imaged by seismic-reflection data decreases slightly with depth. It seems equally likely, however, that these events are associated with the northwest-trending folds and faults here considered to be related to the Devils Mountain fault.

Although fewer earthquakes have been recorded south of the Devils Mountain fault in the study area, at least four events (M>2.5) may correlate with the Utsalady Point or Strawberry Point faults (fig. 2). One M=5.0 event (1/26/1957) occurred beneath Oak Harbor within the Utsalady Point fault zone; a second M=5.0 event (1/28/1945) has been located near a projection of the Utsalady Point fault zone a few kilometers east of Camano Island. Relevant smaller events include a M=3.3 (11/1/2000) earthquake that occurred adjacent to the Utsalady Point zone at Strawberry Point and a M=3.5 event (3/13/1992) that occurred in the eastern Strait of Juan de Fuca close to the location where the Strawberry Point and Utsalady Point faults converge and terminate.

Significance for Earthquake Hazards

Geologic and geophysical data indicate that the Devils Mountain fault and the newly recognized Strawberry Point and Utsalady Point faults are active structures that represent potential earthquake sources. Critical source parameters that contribute to earthquake hazard assessment include fault location, length, slip rate, and recurrence interval of large earthquakes (Wells and Coppersmith, 1994; Frankel and others, 1996). Figures 2, 27, and 28 show the locations of these faults in the eastern Strait of Juan de Fuca and northern Puget Lowland. Large earthquakes generated in this source zone could affect broad areas and large populations in the Puget and Fraser Lowlands and on southern Vancouver Island. Urban centers proximal to the fault zone include Victoria (<5 km from the potential earthquake source), Everett (40 km), Bellingham (45 km), Seattle (75 km), and Vancouver (75 km) (fig. 1). Moreover, significant parts of this region are underlain by large deltas and sedimentary basins, areas that are particularly prone to strong ground motion.

Using the relationship between rupture length and moment magnitude (M) defined by Wells and Coppersmith (1994) and assuming that the entire mapped length of the Devils Mountain fault (≈125 km) could rupture in one event, an earthquake of M=7.5 could be generated on this structure. This investigation recognized two possible segment boundaries on the Devils Mountain fault that might limit rupture during a large earthquake; however, no data are available to test this segmentation hypothesis.

Based on their minimum lengths (≈22–28 km), earthquakes of M≈6.7 could occur on the Strawberry Point and Utsalady Point faults. Each of these faults breaks into several splays east of Whidbey Island (figs. 27, 28), but this bifurcation is unlikely to limit earthquake rupture—it should be assumed that one or more of these splays merges with a single trace of each fault at depth. The Strawberry Point and Utsalady Point faults might also connect at depth, a scenario in which surface deformation could occur along each fault in the same earthquake.

Because incorporating slip rates on faults is an important component of seismic hazard assessment and mapping (Frankel and others, 1996, among others), we presented and explained our best information for each fault (fig. 38). We consider these variably speculative rates as starting points and expect that they will be refined and modified in the future as more data accumulate. As stressed previously ("Fault Mapping, Style, and Rates of Deformation"), determining slip rates across the zone is problematic because (1) data constrain only the vertical component of slip on oblique-slip faults, and (2) both the sense of vertical slip and the vertical slip rate vary along the strike of the Strawberry Point and Utsalady Point faults. Our best estimate for vertical slip on the Devils Mountain fault is 0.05–0.31 mm/yr (fig. 38). We suspect that the lateral component of slip may be larger, but probably no more than about 0.75 mm/yr based on regional geologic considerations. We think reasonable minimum vertical slip rates for both the Strawberry Point and Utsalady Point faults are about 0.2 mm/yr and that field evidence for contractional deformation suggests an additional significant lateral component of slip. Thus, the cumulative slip rate on the three main faults of this complex northern Puget Lowland structural zone probably exceeds 0.5 mm/yr and could be much larger. By comparison, Johnson and others (1999) suggested about 0.7–1.1 mm/yr of slip on faults of the Seattle fault zone.

Paleoseismological investigations of the Devils Mountain fault, Strawberry Point, and Utsalady Point faults are lacking; hence no information exists on earthquake recurrence interval. Williams and Hutchinson (2000) reported evidence for four late Holocene tsunami deposits from a marsh near Swantown on the northwest coast of Whidbey Island (fig. 27). The overlap in age between the two youngest deposits (1,160–1,350 and 1,400–1,700 ^{14}C yr B.P.) and inferred great earthquakes at the Cascadia plate boundary, about 250 km to the west, suggests that they were emplaced by tsunamis from this source area. The two older layers (1,810–2,060 and 1,830–2,020 ^{14}C yr B.P.) do not correlate with plate boundary events and may be products of earthquakes on local faults in the eastern Strait of Juan de Fuca, such as the Devils Mountain, Strawberry Point, and Utsalady Point faults. In another investigation, Bourgeois and Johnson (2001) found evidence of three liquefaction events in post A.D. 900 deltaic deposits from the Snohomish River delta, 35 km south of the Devils Mountain fault. One of these events has been correlated with a strong earthquake on the Seattle fault, but the others have not been correlated with known events; the Devils Mountain, Strawberry Point, and Utsalady Point faults are nearby possible earthquake sources. Thorson (1996) speculated that earthquake cycles and recurrence intervals for Puget Lowland faults may have been perturbed during glacial loading and rebound, but this hypothesis cannot be tested with available data for any fault in the Puget Lowland.

Conclusions

Marine seismic-reflection surveys, aeromagnetic mapping, coastal exposures of Pleistocene strata, and lithologic logs of water wells provide important information for assessing the active tectonics of the northern Puget Lowland and eastern Strait of Juan de Fuca region of the Pacific Northwest. These data indicate that the Devils Mountain fault and the newly recognized Strawberry Point and Utsalady Point faults are active structures and represent potential earthquake sources.

The north-dipping (45°–75°) Devils Mountain fault extends westward for more than 125 km from the Cascade Range foothills to Vancouver Island, where it may merge with the Leech River and (or) San Juan faults. The Devils Mountain fault is bounded by northwest-trending en-echelon folds and faults, a map pattern strongly suggesting that it is a left-lateral, oblique-slip, transpressional "master fault." Aeromagnetic anomalies coincide with both the trace of the Devils Mountain fault and en-echelon structures. Quaternary strata are deformed on nearly all seismic-reflection profiles, and onshore subsurface data suggest offset of upper Pleistocene strata.

The west-northwest-trending, subvertical Strawberry Point fault cuts across northern Whidbey Island and has a minimum length of 22 km. On the west coast of Whidbey Island and in the Strait of Juan de Fuca, the fault has south-side-up offset and forms the northern boundary of an uplift of pre-Tertiary basement rock. The fault bifurcates into a 2-km-wide zone as it crosses Whidbey Island. A stratigraphic correlation diagram based on outcrops and water-well logs suggests that each of the four fault splays within this zone has apparent north-side-up offset. Outcrops at Strawberry Point also indicate considerable shortening (dips as steep as 45°) in upper Pleistocene strata between the faults. The vertical fault trace, reversal of offset, and evidence for associated contractional deformation indicate that the Strawberry Point fault is also an oblique-slip, transpressional fault.

The northwest-trending, subvertical Utsalady Point fault similarly cuts across northern Whidbey Island and has a minimum length of 28 km. It forms the southern margin of the uplifted pre-Tertiary basement block on the west coast of Whidbey Island, where it has north-side-up offset. Offshore seismic-reflection data from east of Whidbey Island indicate that it also bifurcates into a broad (1.5 km) zone of several splays to the east. Onshore outcrops and subsurface logs from Camano Island indicate a probable reversal of offset (to south side up) along the zone; they also reveal both faulting and folding (dips as steep as 24°) in upper Pleistocene strata. As with the Strawberry Point fault, the vertical fault trace(s), reversal of offset, and evidence for associated contractional deformation suggest that the Utsalady Point fault is an oblique-slip, transpressional fault.

Collectively, the Devils Mountain, Strawberry Point, and Utsalady Point faults represent an active complex, distributed, transpressional deformation zone. The cumulative slip rate on three main faults of this zone probably exceeds 0.5 mm/yr and could be much larger.

References Cited

Aber, J.S., Croot, D.G., and Fenton, M.M., 1989, Glaciotectonic landforms and structures: Dordrecht, Kluwer Academic Publishers, 200 p.

Adair, M.J., Talmage, R.H., Crosby, T.W., and Testa, S.M., 1989, Geology and seismicity of the Skagit nuclear power plant site: Washington Division of Geology and Earth Resources Bulletin 78, p. 607–624.

Anderson, H.W., Jr., 1968, Groundwater resources of Island County: Washington Division of Water Resources, Water Supply Bulletin 25, Part II, 317 p.

Armstrong, J.E., Crandell, D.R., Easterbrook, D.J., and Noble, J.B., 1965, Late Pleistocene stratigraphy and chronology in southwestern British Columbia and northwestern Washington: Geological Society of America Bulletin, v. 76, p. 321–330.

Barnes, P.M., and Audru, J.C., 1999, Recognition of active strike-slip faulting from high-resolution marine seismic reflection profiles—Eastern Marlborough fault system, New Zealand: Geological Society of America Bulletin, v. 111, p. 538–559.

Berger, G.W., and Easterbrook, D.J., 1993, Thermoluminesence dating tests for lacustrine, glaciomarine, and floodplain sediments from western Washington and British Columbia: Canadian Journal of Earth Sciences, v. 30, p. 1815–1828.

Blakely, R.J., and Lowe, Carmel, 2001, Aeromagnetic anomalies of the eastern Strait of Juan de Fuca region, *in* Mosher, D.C., and Johnson, S.Y., eds., Rathwell, G.J., Kung, R.B., and Rhea, S.B., compilers, Neotectonics of the eastern Strait of Juan de Fuca; a digital geological and geophysical atlas: Geological Survey of Canada Open File Report D3931 (CD digital product), 2 maps and text.

Blakely, R.J., Parsons, T.E., Brocher, T.M., Langenheim, V.E., and ten Brink, Uri, 1999, A three-dimensional view of the Seattle basin, Washington, from gravity inversion and seismic velocity [abs.]: Eos, v. 80, p. F762.

Blakely, R.J., Wells, R.E., and Weaver, C.S., 1999, Puget Sound aeromagnetic maps and data: U.S. Geological Survey Open-File Report 99-514.

Blunt, D.J., Easterbrook, D.J., and Rutter, N.W., 1987, Chronology of Pleistocene sediments in the Puget Lowland, Washington: Washington Division of Geology and Earth Resources Bulletin, v. 77, p. 321–353.

Bohannon, R.G., Gardner, J.V., Sliter, R., and Normark, W., 1998, Seismic hazard potential of offshore Los Angeles basin based on high-resolution, multibeam bathymetry and close-spaced seismic-reflection profiles [abs.]: Eos, v. 79, p. F818.

Booth, D.B., 1994, Glaciofluvial infilling and scour of the Puget Lowland, Washington, during ice-sheet glaciation: Geology, v. 22, p. 695–698.

Bourgeois, Joanne, and Johnson, S.Y., 2001, Geologic evidence of earthquakes at the Snohomish delta: Geological Society of America Bulletin, v. 113, p. 482–494.

Brandon, M.T., Cowan, D.S., and Vance, J.A., 1988, The Late Cretaceous San Juan thrust system, San Juan Islands, Washington: Geological Society of America Special Paper 221, 81 p.

Brocher, T.M., and Ruebel, A.L., 1998, Compilation of 29 sonic and density logs from 23 oil test wells in western Washington State: U.S. Geological Survey Open-File Report 98-249, 60 p.

Bucknam, R.C., Hemphill-Haley, E., and Leopold, E.B., 1992, Abrupt uplift within the past 1,700 years at southern Puget Sound, Washington: Science, v. 258, p. 1611–1614.

Bucknam, R.C., Sherrod, B.L., and Elfendahl, G., 1999, A fault scarp of probable Holocene age in the Seattle fault zone: Seismological Research Letters, v. 70, p. 233.

Christie-Blick, Nicholas, and Biddle, K.T., 1985, Deformation and basin formation along strike-slip faults, *in* Biddle, K.T., and Christie-Blick, Nicholas, eds., Strike-slip deformation, basin formation, and sedimentation: Society of Economic Paleontologists and Mineralogists Special Publication 37, p. 1–34.

Clague, J.J., 1978, Mid Wisconsinan climates of the Pacific Northwest, *in* Current Research, Part B: Geological Survey of Canada Paper 78-1B, p. 95-100.

———1994, Quaternary stratigraphy and history of south-coastal British Columbia, *in* Monger, J.W.H., ed., Geology and geological hazards of the Vancouver region, southwestern British Columbia: Geological Survey of Canada Bulletin 481, p. 181–192.

Clague, J.J., Harper, J.R., Hebda, R.J., and Howes, D.E., 1982, Late Quaternary sea levels and crustal movements, coastal British Columbia: Canadian Journal of Earth Sciences, v. 19, p. 597–618.

Cline, D.R., Jones, M.A., Dion, N.P., Whiteman, K.J., and Sapik, D.B., 1982, Preliminary survey of ground-water resources for Island County, Washington: U.S. Geological Survey Water Resources Investigation Open-File Report 82-561, 46 p.

Clowes, R.M., Brandon, M.T., Green, A.G., Yorath, C.J., Sutherland Brown, A., Kanesewich, E.R., and Spencer, C.J., 1987, LITHOPROBE—southern Vancouver Island—Cenozoic subduction complex imaged by deep seismic reflections: Canadian Journal of Earth Sciences, v. 24, p. 31–51.

Cooke, S.S., ed., 1997, A field guide to the common wetland plants of western Washington and northwestern Oregon: Seattle, Wash., Seattle Audubon Society, 415 p.

Davies, T.A., Bell, Trevor, Cooper, A.K., Josenhans, Heiner, Polyak, Leonid, Solheim, Anders, Stoker, M.S., and Stravers, J.A., eds., 1997, Glaciated continental margins—An atlas of acoustic images: London, Chapman and Hall, 315 p.

Dethier, D.P., Pessl, Fred, Jr., Keuler, R.F., Balzarini, M.A., and Pevear, D.R., 1995, Late Wisconsin glaciomarine deposition and isostatic rebound, northern Puget Lowland, Washington: Geological Society of America Bulletin, v. 107, p. 1288–1303.

Dragert, Herbert, and Hyndman, R.D., 1995, Continuous GPS monitoring of elastic strain in the northern Cascadia subduction zone: Geophysical Research Letters, v. 22, p. 755–758.

Dragovich, J.D., and Grisamer, C.L., 1998, Quaternary stratigraphy, cross sections, and general geohydrologic potential of the Bow and Alger 7.5-minute quadrangles, western Skagit County, Washington: Washington Division of Geology and Earth Resources Open-File Report 98-8, 30 p., 6 plates.

Dragovich, J.D., Norman, D.K., Grisamer, C.L., Logan, R.L., and Anderson, Garth, 1998, Geologic map and interpreted geologic history of the Bow and Alger 7.5-minute quadrangles, western Skagit County, Washington: Washington Division of Geology and Earth Resources Open-File Report 98-5, 80 p., 3 plates.

Dragovich, J.D., Norman, D.K., Troost, M.L., Anderson, Garth, and McKay, D.T., 2000, Geologic map of the Anacortes South and LaConner 7.5-minute quadrangles, western Skagit county, Washington: Washington Division of Geology and Earth Resources Open-File Report 2000-6, 4 plates, scale 1:24,000.

Easterbrook, D.J., 1968, Pleistocene stratigraphy of Island County: Washington Division of Water Resources, Water Supply Bulletin 25, Part I, 34 p.

———1969, Pleistocene chronology of the Puget Lowland and San Juan Islands, Washington: Geological Society of America Bulletin, v. 80, p. 2273–2286.

———1976, Stratigraphy and palynology of late Quaternary sediments in the Puget Lowland, Washington—Reply: Geological Society of America Bulletin, v. 87, p. 155–156.

———1994a, Chronology of pre-late Wisconsin Pleistocene sediments in the Puget Lowland, Washington, *in* Lasmanis, R., and Cheney, E.S., eds., Regional geology of Washington State: Washington Division of Geology and Earth Resources Bulletin 80, p. 191–206.

———1994b, Stratigraphy and chronology of early to late Pleistocene glacial and interglacial sediments in the Puget Lowland, Washington, *in* Swanson, D.A., and Haugerud, R.A., eds., Geologic field trips in the Pacific Northwest (published for the 1994 Annual Meeting of the Geological Society of America): Seattle, Wash., University of Washington, p. 1J-1-38.

Easterbrook, D.J., Crandell, D.R., and Leopold, E.B., 1967, Pre-Olympia Pleistocene stratigraphy and chronology in the central Puget Lowland, Washington: Geological Society of America Bulletin, v. 78, p. 13–20.

Engebretson, D.C., Cox, Allan, and Gordon, R.G., 1985, Relative motions between oceanic and continental plates of the Pacific Basin: Geological Society of America Special Paper 206, 59 p.

Eronen, Matti, Kankainen, Tuovi, and Tsukada, Matsuo, 1987, Late Holocene sea-level record in a core from the Puget Lowland, Washington: Quaternary Research, v. 27, p. 147–159.

Evans, J.E., and Ristow, John, Jr., 1994, Depositional history of the southeastern outcrop belt of the Chuckanut Formation—Implications for the Darrington-Devils Mountain and Straight Creek fault zones, Washington (U.S.A.): Canadian Journal of Earth Sciences, v. 31, p. 1727–1743.

Fader, G.B.J., 1997, Effects of shallow gas on seismic-reflection profiles, *in* Davies, T.A., Bell, Trevor, Cooper, A.K., Josenhans, Heiner, Polyak, Leonid, Solheim, Anders, Stoker, M.S., and Stravers, J.A., eds., Glaciated continental margins—An atlas of acoustic images: London, Chapman and Hall, p. 29–30.

Finn, Carol, Phillips, W.M., and Williams, D.L., 1991, Gravity anomaly and terrain maps of Washington: U.S. Geological Survey Geophysical Investigations Map GP-988, scale 1:500,000 and 1:1,000,000.

Fisher, M.A., and Expedition Participants, 1999, Seismic survey probes urban earthquake hazards in Pacific Northwest: Eos, v. 80, p. 13–17.

Frankel, A.D., Mueller, C.S., Barnhard, T.A., Perkins, D.M., Leyendecker, E.F., Dickman, N.C., Hanson, S.L., and Hopper, M.G., 1996, National seismic hazards maps, June 1996 documentation: U.S. Geological Survey Open-File Report 96-532.

Fulton, R.J., Armstrong, J.E., and Fyles, J.G., 1976, Stratigraphy and palynology of late Quaternary sediments in the Puget Lowland, Washington—Discussion: Geological Society of America Bulletin, v. 87, p. 153–155.

Gower, H.D., Yount, J.C., and Crosson, R.S., 1985, Seismotectonic map of the Puget Sound region, Washington: U.S. Geological Survey Miscellaneous Investigations Series Map I-1613, scale 1:250,000.

Gradstein, F.M., and Ogg, J.G., 1996, A Phanerozoic time scale: Episodes, v. 19, p. 3–4.

Hansen, B.S., and Easterbrook, D.J., 1974, Stratigraphy and palynology of late Quaternary sediments in the Puget Lowland, Washington: Geological Society of America Bulletin, v. 85, p. 587–602.

Harding, D.J., and Berghoff, G.S., 2000, Fault scarp detection beneath dense vegetation cover—Airborne laser mapping of the Seattle fault zone, Bainbridge Island, Washington State: Proceedings of the American Society of Photogrammetry and Remote Sensing Annual Conference, Washington, D.C., 9 p.

Harding, T.P., Gregory, R.F., and Stephens, L.H., 1983, Convergent wrench fault and positive flower structure, Ardmore Basin, Oklahoma, *in* Bally, A.W., ed., Seismic expression of structural styles: American Association of Petroleum Geologists Studies in Geology Series 15, v. 3, p. 4.2-13-17.

Hart, J.K., and Boulton, G.S., 1991, The interrelation of glaciotectonic and glaciodepositional processes within the glacial environment: Quaternary Science Reviews, v. 10, p. 335–350.

Hart, J.K., and Roberts, D.H., 1994, Criteria to distinguish between subglacial glaciotectonic and glaciomarine sedimentation—I, Deformation styles and sedimentology: Sedimentary Geology, v. 91, p. 191–213.

Heusser, C.J., and Heusser, L.E., 1981, Palynology and paleotemperature of the Whidbey Formation, Puget Lowland, Washington: Canadian Journal of Earth Sciences, v. 18, p. 136–149.

Hewitt, A.T., and Mosher, D.C., 2001, Surficial geology of the eastern Juan de Fuca Strait, *in* Mosher, D.C., and Johnson, S.Y., eds., Rathwell, G.J., Kung, R.B., and Rhea, S.B., compilers, Neotectonics of the eastern Strait of Juan de Fuca; a digital geological and geophysical atlas: Geological Survey of Canada Open File Report D3931 (CD digital product), 1 map with text.

Hicock, S.R., and Armstrong, J.E., 1981, Coquitlam Drift — A pre-Vashon Fraser glacial formation in the Fraser Lowland, British Columbia: Canadian Journal of Earth Sciences, v. 18, p. 1443–1451.

Hobbs, S.W., and Pecora, W.T., 1941, Nickel-gold deposit near Mount Vernon, Skagit County, Washington, Chapter D of Strategic minerals investigations, 1941, short papers and preliminary reports, Part 1: U.S. Geological Survey Bulletin 931, p. 57–78.

Hovland, M., and Judd, A.G., 1988, Seabed pockmarks and seepages; impact on geology, biology, and the marine environment: London, Graham and Trotman, Inc., 293 p.

Jervey, M.T., 1988, Quantitative geological modeling of siliciclastic rock sequences and their seismic expression, *in* Wilgus, C.K., Posamentier, H., Ross, C.A., and Kendall, G. St. C., Sea level changes—An integrated approach: Society of Economic Paleontologists and Mineralogists Special Publication 42, p. 47–70.

Johnson, S.Y., 1984a, Stratigraphy, age, and paleogeography of the Eocene Chuckanut Formation, northwest Washington: Canadian Journal of Earth Sciences, v. 21, p. 92–106.

———1984b, Evidence for a margin-truncating transcurrent fault (pre-Late Eocene) in western Washington: Geology, v. 12, p. 538–541.

Johnson, S.Y., Dadisman, S.V., Childs, J.R., and Stanley, W.D., 1999, Active tectonics of the Seattle fault and central Puget Lowland—Implications for earthquake hazards: Geological Society of America Bulletin, v. 111, p. 1042–1053, and oversize insert.

Johnson, S.Y., Mosher, D.C., Dadisman, S.V., Childs, J.R., and Rhea, S.B., 2001, Tertiary and Quaternary structures of the eastern Strait of Juan de Fuca—Interpreted map, *in* Mosher, D.C., and Johnson, S.Y., eds., Rathwell, G.J., Kung, R.B., and Rhea, S.B., compilers, Neotectonics of the eastern Strait of Juan de Fuca; a digital geological and geophysical atlas: Geological Survey of Canada Open File Report D3931 (CD digital product), 1 map with text and figures.

Johnson, S.Y., Potter, C.J., and Armentrout, J.M., 1994, Origin and evolution of the Seattle basin and Seattle fault: Geology, v. 22, p. 71–74 and oversize insert.

Johnson, S.Y., Potter, C.J., Armentrout, J.M., Miller, J.J., Finn, Carol, and Weaver, C.S., 1996, The southern Whidbey Island fault, an active structure in the Puget Lowland, Washington: Geological Society of America Bulletin, v. 108, p. 334–354 and oversize insert.

Jones, M.A., 1985, Occurrence of groundwater and potential for seawater intrusion, Island County, Washington: U.S. Geological Survey Water Investigations Report 85-4046, 6 sheets.

———1996, Thickness of unconsolidated deposits in the Puget Sound Lowland, Washington and British Columbia: U.S. Geological Survey Water Resources Investigations Report 94-4133.

Khazaradze, Giorgi, Qamar, Anthony, and Dragert, Herbert, 1999, Tectonic deformation in western Washington from continuous GPS measurements: Geophysical Research Letters, v. 26, p. 3153–3156.

Kitsap County Public Utility District, 1998, Bainbridge Island, Washington (Lidar shaded-relief map): Kitsap Public Utility District Geographic Information System map, scale 1:18,000.

Kukla, George, McManus, J.F., Rousseau, D.D., and Chuine, Isabelle, 1997, How long and how stable was the last interglacial: Quaternary Science Reviews, v. 16, p. 605–612.

Loveseth, T.P., 1975, The Devils Mountain fault zone, northwestern Washington: Seattle, Wash., University of Washington M.S. thesis, 29 p.

Ludwin, R.S., and Qamar, A.I., 1995, Historic seismicity catalog and macroseismic accounts for Cascadia, 1793–1929: Seattle, Wash., University of Washington Geophysics Program, 72 p.

MacLeod, N.S., Tiffin, D.L., Snavely, P.D., Jr., and Currie, R.G., 1977, Geologic interpretation of magnetic and gravity anomalies in the Strait of Juan de Fuca, U.S.-Canada: Canadian Journal of Earth Sciences, v. 14, p. 223–238.

Marcus, K.L., 1980, Eocene-Oligocene sedimentation and deformation in the northern Puget Sound area, Washington: Northwest Science, v. 9, p. 52–58.

McManus, J.F., Bond, G.C., Broecker, W.S., Johnsen, S., Labeyrle, L., and Higgins, S., 1994, High-resolution climate records from the north Atlantic during the last interglacial: Nature, v. 371, p. 326–329.

Mosher, D.C., and Simpkin, P.G., 1999, Environmental marine geoscience 1, Status and trends of marine high-resolution seismic reflection profiling—Data acquisition: Geoscience Canada, v. 26, p. 174–188

Mosher, D.C., Kung, R.B., and Hewitt, A.T., 2001, Seafloor morphology of the eastern Juan de Fuca Strait, *in* Mosher, D.C., and Johnson, S.Y., eds., Rathwell, G.J., Kung, R.B., and Rhea, S.B., compilers, Neotectonics of the eastern Strait of Juan de Fuca; a digital geological and geophysical atlas: Geological Survey of Canada Open File Report D3931 (CD digital product), 1 map with text and figures.

Muhs, D.R., Kennedy, G.L., and Rockwell, T.K., 1994, Uranium-series ages of marine terrace corals from the Pacific Coast of North America and implications for last-interglacial sea level history: Quaternary Research, v. 42, p. 72–87.

Muller, J.E., 1983, Geology, Victoria: Geological Survey of Canada Map 1553A, scale 1:100,000.

Naugler, W.E., Karlin, R.E., and Holmes, M.L., 1996, Lakes as windows in the crust—Lake Cavanaugh and the Devils Mountain fault: Geological Society of America Abstracts with Programs, v. 28, p. 95.

Nelson, A.R., Pezzopane, S.K., Bucknam, R.C., Koehler, R., Narwold, C., Kelsey, H.M., LaPrade, W.T., Wells, R.E., and Johnson, S.Y., 1999, Late Holocene surface faulting in the Seattle fault zone on Bainbridge Island, Washington: Seismological Research Letters, v. 70, p. 233.

Pessl, Fred, Jr., Dethier, D.P., Booth, D.B., and Minard, J.P., 1989, Surficial geologic map of the Port Townsend 30- by 60- minute quadrangle, Puget Sound region, Washington: U.S. Geological Survey Miscellaneous Investigations Series Map I-1198-F, scale 1:100,000.

Pillans, B.J., Chappell, John, and Naish, T.R., 1998, A review of the Milankovitch climatic beat—Template for Plio-Pleistocene sea-level changes and sequence stratigraphy: Sedimentary Geology, v. 122, p. 5–21.

Porter, S.C., and Swanson, T.W., 1998, Radiocarbon age constraints on advance and retreat of the Puget Lobe of the Cordilleran ice sheet during the last glaciation: Quaternary Research, v. 50, p. 205–213.

Posamentier, H.W., Jervey, M.T., and Vail, P.R., 1988, Eustatic controls on clastic deposition I—Conceptual framework, *in* Wilgus, C.K., Posamentier, H.W., Ross, C.A., and Kendall, G.St.C., Sea level changes—An integrated approach: Society of Economic Paleontologists and Mineralogists Special Publication 42, p. 109–124.

Pratt, T.L., Johnson, S.Y., Potter, C.J., and Stephenson, W.J., 1997, Seismic reflection images beneath Puget Sound, western Washington State—The Puget Lowland thrust sheet hypothesis: Journal of Geophysical Research, v. 102, p. 27469–27490.

Puget Power, 1979, Preliminary safety analysis report—Skagit Nuclear Power Project: Seattle, Wash., Puget Power, Chapter 2, Appendix G.

Rau, W.W., and Johnson, S.Y., 1999, Well stratigraphy and correlations, western Washington and northwest Oregon: U.S. Geological Survey Geologic Investigations Series Map I-2621, 3 oversized sheets/charts, 31 p.

Rockwell, T.K., Keller, E.A., and Dembroff, G.R., 1988, Quaternary rate of folding of the Ventura anticline, western Transverse Ranges, southern California: Geological Society of America Bulletin, v. 100, p. 850–858.

Roddick, J.A., Muller, J.E., and Okulitch, A.V., 1979, Fraser River, British Columbia-Washington: Geological Survey of Canada Map 1386A, Sheet 92, scale 1:1,000,000.

Sangree, J.B., and Widmier, J.M., 1977, Seismic stratigraphy and global changes in sea level, part 9—Seismic interpretation of clastic depositional facies, *in* Payton, C.E., ed., Seismic stratigraphy—Applications to hydrocarbon exploration: American Association of Petroleum Geologists Memoir 26, p. 165–184.

Sapik, D.B., Bortleson, G.C., Drost, B.W., Jones, M.A., and Frych, E.A., 1988, Ground-water resources and simulation of flow in aquifers containing freshwater and seawater, Island County, Washington: U.S. Geological Survey Water Investigations Report 87-4182, 67 p.

Schneider, C.L., Hummon, C., Yeats, R.S., and Huftile, G.L., 1996, Structural evolution of the northern Los Angeles basin, California, based on growth strata: Tectonics, v. 15, p. 341–355.

Shackleton, N.J., Imbrie, John, and Hall, M.A., 1983, Oxygen and carbon isotope record of east Pacific core V19-30—Implications for the formation of deep water in the late Pleistocene North Atlantic: Earth and Planetary Science Letters, v. 65, p. 233–244.

Spiegel, M.R., 1961, Schaum's outline of theory and problems of statistics: New York, Schaum Publishing Company, 359 p.

Stanley, W.D., Villasenor, Antonio, and Benz, H.M., 1999, Subduction and crustal dynamics of western Washington—A tectonic model for earthquake hazard evaluation: U.S. Geological Survey Open-File Report 99-311, 90 p.

Stoker, M.S., Pheasant, J.B., and Josenhans, Heiner, 1997, Seismic methods and interpretation, *in* Davies, T.A., Bell, Trevor, Cooper, A.K., Josenhans, Heiner, Polyak, Leonid, Solheim, Anders, Stoker, M.S., and Stravers, J.A., eds., Glaciated continental margins—An atlas of acoustic images: London, Chapman and Hall, p. 9–26.

Stoffel, K.L., 1981, Stratigraphy of pre-Vashon Quaternary sediments applied to the evaluation of a proposed major structure in Island County, Washington: U.S. Geological Survey Open-File Report 81-292, 155 p.

Suppe, John, and Medwedeff, D.A., 1990, Geometry and kinematics of fault-propagation folding: Ecologae Geologicae Helvetiae, v. 83, p. 409–454.

Tabor, R.W., 1994, Late Mesozoic and possible early Tertiary accretion in western Washington state—The Helena-Haystack melange and the Darrington-Devils Mountain fault zone: Geological Society of America Bulletin, v. 106, p. 217–232.

Tabor, R.W., Booth, D.B., Vance, J.A., and Ford, A.B., 1988, Geologic map of the Sauk River 30- by 60-minute quadrangle, Washington: U.S. Geological Survey Open-File Report 88-692, scale 1:100,000.

Tabor, R.W., and Cady, W.M., 1978, Geologic map of the Olympic Peninsula, Washington: U.S. Geological Survey Miscellaneous Investigations Series Map I-994, scale 1:125,000.

Tabor, R.W., Frizzell, V.A., Jr., Booth, D.B., Waitt, R.B., Jr., Whetten, J.T., and Zartman, R.E., 1993, Geologic map of the Skykomish River 30- by 60-minute quadrangle, Washington: U.S. Geological Survey Miscellaneous Investigations Series Map I-1963, scale 1:100,000.

Thorson, R.M., 1996, Earthquake recurrence and glacial loading in western Washington: Geological Society of America Bulletin, v. 108, p. 1182–1191.

Troost, K.G., 1999, The Olympia nonglacial interval in the southcentral Puget Lowland, Washington: Seattle, Wash., University of Washington M.S. thesis, 122 p.

Walsh, T.J., Korosec, M.A., Phillips, W.M., Logan, F.J., and Schasse, H.W., 1987, Geologic map of Washington—Southwest quadrant: Washington Division of Geology and Earth Resources, Geologic Map GM-34, scale 1:250,000.

Wang, Kelin, 1996, Simplified analysis of horizontal stresses in a buttressed forearc sliver at an oblique subduction zone: Geophysical Research Letters, v. 23, p. 2012–2024.

Washington Public Power Supply System (WPPSS), 1981, Regional geologic map, *in* WPPSS nuclear project no. 2—Final safety analysis report, Amendment 18: WPPSS Docket no. 50-397, v. 2, scale 1:1,000,000.

Wells, D.L., and Coppersmith, K.J., 1994, New empirical relationships among magnitude, rupture length, rupture width, rupture area, and surface displacement: Bulletin of the Seismological Society of America, v. 84, p. 974–1002.

Wells, R.E., Weaver, C.S., and Blakely, R.J., 1998, Fore-arc migration in Cascadia and its neotectonic significance: Geology, v. 26, p. 759–762.

Whetten, J.T., 1978, The Devils Mountain fault—A major Tertiary structure in northwest Washington: Geological Society of America Abstracts with Programs, v. 10, p. 153.

Whetten, J.T., Carroll, P.I., Gower, H.D., Brown, E.H., and Pessl, Fred, Jr., 1988, Bedrock geologic map of the Port Townsend 30- by 60-minute quadrangle, Puget Sound Region, Washington: U.S. Geological Survey Miscellaneous Investigations Series Map I-1198-G, scale 1:100,000.

Whetten, J.T., Zartman, R.E., Blakely, R.J., and Jones, D.L., 1980, Allochthonous Jurassic ophiolite in northwest Washington: Geological Society of America Bulletin, v. 91, p. 359–368.

Williams, Harry, and Hutchinson, Ian, 2000, Stratigraphic and microfossil evidence for late Holocene tsunamis at Swantown marsh, Whidbey Island, Washington: Quaternary Research, v. 54, p. 218–227.

Yount, J.C., Dembroff, G.R., and Barats, G.M., 1985, Map showing depth to bedrock in the Seattle 30' by 60' quadrangle, Washington: U.S. Geological Survey Miscellaneous Field Studies Map MF-1692, scale 1:100,000.

Yount, J.C., and Gower, H.D., 1991, Bedrock geologic map of the Seattle 30' by 60' quadrangle, Washington: U.S. Geological Survey Open-File Report 91-147, 37 p., 4 plates.

Published in the Central Region, Denver, Colorado
Manuscript approved for publication April 10, 2000
Graphics by S.Y. Johnson
Photocomposition by Norma J. Maes
Edited by L.M. Carter

☆ U.S. GOVERNMENT PRINTING OFFICE: 2001 — 673-049 / 30052 Region No. 8